UNSUPERVISED ADAPTIVE FILTERING

Volume II: Blind Deconvolution

ADAPTIVE AND LEARNING SYSTEMS FOR SIGNAL PROCESSING, COMMUNICATIONS, AND CONTROL

Editor: Simon Haykin

Werbos / THE ROOTS OF BACKPROPAGATION: From Ordered Derivatives to Neural Networks and Political Forecasting

Krstić, Kanellakopoulos, and Kokotović / NONLINEAR AND ADAPTIVE CONTROL DESIGN

Nikias and Shao / SIGNAL PROCESSING WITH ALPHA-STABLE DISTRIBUTIONS AND APPLICATIONS

Diamantaras and Kung / PRINCIPAL COMPONENT NEURAL NETWORKS: THEORY AND APPLICATIONS

Tao and Kokotović / ADAPTIVE CONTROL OF SYSTEMS WITH ACTUATOR AND SENSOR NONLINEARITIES

Tsoukalas / FUZZY AND NEURAL APPROACHES IN ENGINEERING

Hrycej / NEUROCONTROL: TOWARDS AN INDUSTRIAL CONTROL METHODOLOGY

Beckerman / ADAPTIVE COOPERATIVE SYSTEMS

Cherkassky and Mulier / LEARNING FROM DATA: CONCEPTS, THEORY, AND METHODS

Passino and Burgess / STABILITY OF DISCRETE EVENT SYSTEMS

Sánchez-Peña and Sznaier / ROBUST SYSTEMS THEORY AND APPLICATIONS

Vapnik / STATISTICAL LEARNING THEORY

Haykin and Puthusserypady / CHAOTIC DYNAMICS OF SEA CLUTTER

Haykin (ed.) / UNSUPERVISED ADAPTIVE FILTERING, VOLUME I: BLIND SOURCE SEPARATION

Haykin (ed.) / UNSUPERVISED ADAPTIVE FILTERING, VOLUME II: BLIND DECONVOLUTION

UNSUPERVISED ADAPTIVE FILTERING

Volume II: Blind Deconvolution

EDITED BY

Simon Haykin

McMaster University

A WILEY-INTERSCIENCE PUBLICATION

JOHN WILEY & SONS, INC.

New York / Chichester / Weinheim / Brisbane / Singapore / Toronto

This book is printed on acid-free paper. ♾

Copyright © 2000 by John Wiley & Sons, Inc. All rights reserved

Published simultaneously in Canada.

No part of this publication may be reproduced, stored in a retrieval system or transmitted in any form or by any means, electronic, mechanical, photocopying, recording, scanning or otherwise, except as permitted under Section 107 or 108 of the 1976 United States Copyright Act, without either the prior written permission of the Publisher, or authorization through payment of the appropriate per-copy fee to the Copyright Clearance Center, 222 Rosewood Drive, Danvers, MA 01923, (978) 750-8400, fax (978) 750-4744. Requests to the Publisher for permission should be addressed to the Permissions Department, John Wiley & Sons, Inc., 605 Third Avenue, New York, NY 10158-0012, (212) 850-6011, fax (212) 850-6008. E-Mail: PERMREQ@WILEY.COM.

For ordering and customer service, call 1-800-CALL WILEY.

Library of Congress Cataloging-in-Publication Data:

Unsupervised adaptive filtering / Simon Haykin, ed.
p. cm.
"A Wiley-Interscience publication."
Includes bibliographical references and index.
ISBN 0-471-37941-7 (alk. paper)
1. Adaptive filters. 2. Adaptive signal processing. I. Haykin, Simon S., 1931–

TK7872.F5 U57 2000
621.3815′324—dc21 99-029980

CONTENTS

CONTRIBUTORS

James D. Behm
9612 Eagle Court
Ellicott City, Maryland, USA
e-mail: jimbehm@bayserve.net

D. Rick Brown
School of Electrical Engineering
Cornell University
Ithaca, New York, USA
e-mail: {schniter,endres,brownd,raulc}@ee.cornell.edu

Raul A. Casas
School of Electrical Engineering
Cornell University
Ithaca, New York, USA
e-mail: {schniter,endres,brownd,raulc}@ee.cornell.edu

Scott C. Douglas
Department of Electrical Engineering
School of Engineering and Applied Science
Southern Methodist University
Dallas, Texas, USA
e-mail: douglas@seas.smu.edu

Thomas J. Endres
School of Electrical Engineering
Cornell University
Ithaca, New York, USA
e-mail: {schniter,endres,brownd,raulc}@ee.cornell.edu

Inbar Fijalkow
Equipe du Traitement des Images et du Signal
ENSEA—UCP, Cergy-Pontoise, France
e-mail: {fijalkow,touzni}@ensea.fr

Michael Green
Department of Engineering
Australian National University
Canberra, Australia
e-mail: michael.green@anu.edu.au

Simon Haykin
Electrical and Computer Engineering
Communications Research Laboratory
McMaster University
Hamilton, Ontario, Canada
e-mail: haykin@mcmaster.ca

C. Richard Johnson, Jr.
School of Electrical Engineering
Cornell University
Ithaca, New York, USA
e-mail: johnson@anise.ee.cornell.edu

S. Lambotharan
School of Electrical and Electronic Engineering
Imperial College of Science, Technology and Medicine
London, United Kingdom
e-mail: lambo@ee.ic.ac.uk

M. G. Larimore
Applied Signal Technology, Inc.
400 W. California Avenue
Sunnyvale, California, USA
e-mail: mgl@appsig.com

Constantinos B. Papadias
Lucent Technologies
Crawford Hill Laboratory
Holmdel, New Jersey, USA
e-mail: papadias@bell-labs.com

Philip Schniter
School of Electrical Engineering
Cornell University
Ithaca, New York
e-mail: {schniter,endres,brownd,raulc}@ee.cornell.edu

Lang Tong
School of Electrical and Systems Engineering
University of Connecticut
Storrs, Connecticut, USA
e-mail: ltong@eng2.uconn.edu

Azzedine Touzni
Equipe du Traitement des Images et du Signal
ENSEA—UCP, Cergy-Pontoise, France
e-mail: {fijalkow,touzni}@ensea.fr

John R. Treichler
Applied Signal Technology, Inc.
490 West California Avenue
Sunnyvale, California, USA
e-mail: jrt@appsig.com

Hanks H. Zeng
School of Electrical and Systems Engineering
University of Connecticut
Storrs, Connecticut, USA
e-mail: zhy@brc.uconn.edu

PREFACE

In 1994 I edited a book on "blind deconvolution," which presented an account of the various algorithms that had been developed essentially for solving the blind channel-equalization problem. The material presented in that book spanned a period of over 25 years, going back to the pioneering work of Robert Lucky in 1966 on the decision-directed mode of operating the least-mean-square algorithm and that of Y. Sato in 1975 on a blind channel-equalization algorithm that bears his name. These two pioneering contributions were followed by another pioneering contribution to blind channel equalization, namely, the constant-modulus algorithm that was developed independently by Godard in 1980 and Treichler and Agee in 1983. Subsequently, it was recognized that these three blind equalization algorithms are members of the family of Bussgang algorithms

In 1994 Pierre Comon published a paper in a signal-processing journal on "independent component analysis," which was followed by Tony Bell and Terry Sejnowski's 1995 paper in a neural computation journal on the Infomax (or, more precisely, the maximum-entropy) algorithm for blind signal separation. Although, indeed, work on the blind signal-separation problem could be traced to a much earlier paper by J. Herault, C. Jutten, and B. Ans that was published in 1985, it would be fair to say that Pierre Comon's paper and that of Tony Bell and Terry Sejnowski served as catalysts for raising the profile of research interests in blind source separation to the extent that the subject has become a "hot" area with potential applications in a variety of diverse fields.

Despite the fact that blind channel equalization and blind source separation have originated in their own somewhat independent ways, they are in actual fact intimately related to each other. Indeed, they constitute the two pillars of unsupervised adaptive filtering. By bringing them together under the umbrella of this new book, organized in two volumes, not only have we provided an up-to-date treatment of blind signal-separation and blind channel-equalization algorithms and their underlying theoretical formalisms but also opened an avenue for the cross-fertilization of new ideas. Volume I of the book covers blind source-separation algorithms, and Volume II covers blind deconvolution (i.e., blind equalization) and its relationship to blind source separation.

I would like to take this opportunity to express my deep gratitude to each and every one of my coauthors for making the writing of this unique two-volume work a reality.

SIMON HAYKIN

Ancaster, Ontario, Canada
March 2000

UNSUPERVISED ADAPTIVE FILTERING

Volume II: Blind Deconvolution

1

INTRODUCTION

Simon Haykin

1.1 WHY ADAPTIVE FILTERING?

In the signal-processing, communications, and control literature, the term *filter* is commonly used to refer to a device or algorithm that is applied to a set of noisy data in order to extract a prescribed quantity of interest. In much of the work done in the past, optimization of the filter has been based on an objective function or index of performance using second-order statistics. When the filter is linear (i.e., it satisfies the principle of superposition) and all the pertinent statistics are known, the solution is defined by the *Wiener filter* (Haykin 1996). When, however, the filter is required to operate in an environment of unknown statistics or a nonstationary environment, an adaptive filter provides an elegant solution to this more difficult problem. The filter starts from an arbitrary initial condition, knowing nothing about the environment, and proceeds in a step-by-step manner toward an optimum solution.

An *adaptive filter* is formally defined as *a self-designing system* that relies on a recursive algorithm for adjusting its free parameters to operate satisfactorily in an unknown environment.

There are different ways of classifying adaptive filters, depending on the feature of interest. With input–output mapping as the feature to focus on, we can classify adaptive filters into two main groups: linear and nonlinear. *Linear adaptive filters* compute an estimate of a desired response by using a linear combination of the available set of observ-

Unsupervised Adaptive Filtering, Volume II, Edited by Simon Haykin.
ISBN 0-471-37941-7 © 2000 John Wiley & Sons, Inc.

ables applied to the input. This form of input–output mapping is satisfied by having a single layer of computational (processing) units or simply a single computational unit as the output layer. On the other hand, when the input–output mapping is required to be nonlinear, we naturally need to use a *nonlinear adaptive filter*. Typically, nonlinear adaptive filters involve the use of one or more layers of hidden computational units in addition to the output layer (Haykin 1999). As such, nonlinear adaptive filters are capable of tackling more difficult signal-processing tasks than linear adaptive filters.

1.2 SUPERVISED AND UNSUPERVISED FORMS OF ADAPTIVE FILTERING

Adaptive filters may also be classified in another way, depending on whether a desired response is available or not. Specifically, we have the following two classes:

- *Supervised Adaptive Filters.* The algorithms used to design this class of filters assume the availability of a training sequence that specifies the desired response for a certain input signal. In a popular approach to the design of supervised adaptive filters, the desired response is compared against the actual response of the filter, and the resulting error signal is used to adjust the free parameters of the filter. The process of parameter adjustments is continued in a step-by-step manner until a steady-state condition is established. In effect, the information contained in the training sequence about the environment is stored as the design values of the filter's free parameters. Thereafter, the filter is ready for testing data not seen before. The *least-mean-square (LMS) algorithm* invented by Widrow and Hoff (1960) some 40 years ago is an example of supervised adaptive filtering that is simple and yet capable of achieving high performance in a robust manner.
- *Unsupervised Adaptive Filtering.* In this second class of adaptive filters, adjustments to the free parameters of the filter are performed without the need for a desired response. For the filter to perform its function, however, its design includes a set of rules that enable the filter to compute an input-output mapping with specific desirable properties. In the signal-processing literature, unsupervised adaptive filtering is often referred to as *blind adaptation*.

In this two-volume book we study the many facets of unsupervised adaptive filters applied to two important signal-processing tasks: as discussed next, Volume I covers blind source separation and Volume II covers blind deconvolution.

1.3 TWO IMPORTANT UNSUPERVISED SIGNAL-PROCESSING TASKS

1.3.1 Blind Source (Signal) Separation

A filtering problem humans are familiar with is the *cocktail party phenomenon.* We have a remarkable ability to focus on a speaker in the noisy environment of a cocktail party, despite the fact that the speech signal originating from that speaker is buried in an undifferentiated noise background due to other interfering conversations in the room. In the context of unsupervised adaptive filtering a similar filtering problem arises under the umbrella of *blind signal (source) separation*, though it must be said at the outset that the present status of this subject is rather primitive in sophistication compared to the cocktail party phenomenon. To formulate the blind signal-separation problem in its basic form, consider a set of unknown source signals $s_1(t), s_2(t), \ldots, s_m(t)$ that are mutually independent of each other. These signals are linearly mixed in an unknown environment to produce the m-by-1 observation vector (see Fig. 1.1)

$$\mathbf{x}(t) = \mathbf{A}\mathbf{s}(t) \tag{1.1}$$

where

$$\mathbf{s}(t) = [s_1(t), s_2(t), \ldots, s_m(t)]^T$$
$$\mathbf{x}(t) = [x_1(t), x_2(t), \ldots, x_m(t)]^T$$

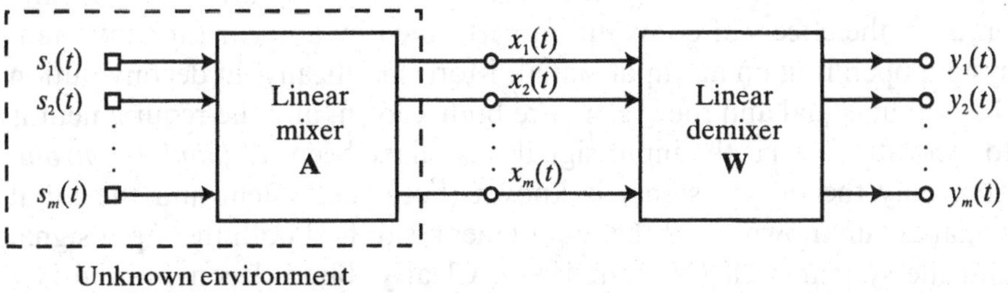

Figure 1.1 Block diagram illustrating the background to the blind signal separation problem in its most basic form.

and $\mathbf{A}$ is an unknown nonsingular *mixing matrix* of dimensions m-by-m. That is, the number of sensors where $\mathbf{x}(t)$ is observed is equal to the number of sources that produce $\mathbf{s}(t)$. Given the observation vector $\mathbf{x}(t)$, the requirement is to recover the original source signals $s_1(t), s_2(t), \ldots, s_m(t)$ in an unsupervised manner. The solution to this problem is feasible, except for an arbitrary *scaling* of each source signal and possible *permutation* of indices under certain but fairly general conditions. In other words, provided the original source signals are independent and the mixing matrix $\mathbf{A}$ is nonsingular, it is possible to find a *demixing matrix* $\mathbf{W}$ defined ideally as follows:

$$\mathbf{y} = \mathbf{W}\mathbf{x} = \mathbf{W}\mathbf{A}\mathbf{x} \rightarrow \mathbf{D}\mathbf{P}\mathbf{s} \tag{1.2}$$

where $\mathbf{y}$ is the output signal vector produced by the demixer, $\mathbf{D}$ is a nonsingular diagonal matrix, and $\mathbf{P}$ is a permutation matrix.

The blind signal-separation problem may be traced to a paper by Herault et al. (1985) published in French. The underlying principle involved in the solution to this problem is nowadays called *independent-components analysis* (ICA) (Comon 1994), which may be viewed as an extension of the widely known principal-components analysis (PCA). Whereas PCA imposes independence in a statistical sense only to the second order while constraining the direction of vectors to be orthogonal, ICA imposes statistical independence on all the individual components of the output vector $\mathbf{y}$, but has no orthogonality constraint. In practice, an algorithmic implementation of ICA can only go for "as statistically independent as possible." In any event, the terms "independent components analysis" and "blind signal (source) separation" mean essentially the same thing; as such, they are used interchangeably in the literature and in this book.

1.3.2 Blind Deconvolution

Turning next to blind deconvolution, the motivation is different. To begin with, *deconvolution* is a signal-processing operation that ideally unravels the effects of convolution performed by a linear time-invariant system operating on an input signal. More specifically, in deconvolution the output signal and the system are both known, and the requirement is to reconstruct what the input signal must have been. In *blind deconvolution*, only the output signal is known (both the system and the input signal are unknown), and the requirement is to find both the input signal and the system itself (Haykin 1994). Clearly, blind deconvolution is a more difficult signal-processing task than ordinary deconvolution.

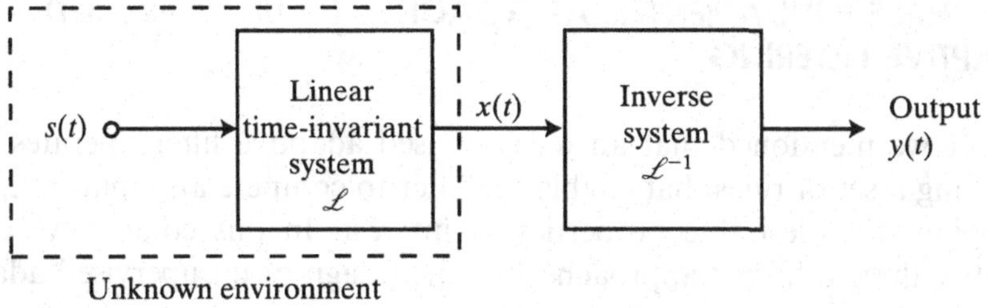

Figure 1.2 Block diagram illustrating the background to the blind deconvolution problem.

To be specific, consider an unknown linear time-invariant system $\mathscr{L}$ with input $s(t)$ assumed to consist of independent and identically distributed symbols. The only thing known about the input is its probability distribution. The requirement is to restore $x(t)$, or equivalently to identify the inverse $\mathscr{L}^{-1}$ of the system $\mathscr{L}$, given the observed signal $x(t)$ at the output of system $\mathscr{L}$, as illustrated in Fig. 1.2.

If the system $\mathscr{L}$, assumed to be a discrete-time system, is *minimum phase* (i.e., the transfer function of the system has all of its poles and zeros confined to the interior of the unit circle in the z-plane), then not only is the system $\mathscr{L}$ stable but so is the inverse system $\mathscr{L}^{-1}$. In this case we can view the unknown input signal $s(t)$ as the "innovation" of the unknown system output $x(t)$, in which case the inverse system $\mathscr{L}^{-1}$ is just a *whitening filter*; with this observation, the blind deconvolution problem is solved.

In many practical situations, however, the system $\mathscr{L}$ may *not* be minimum phase. A discrete-time system is said to be *nonminimum phase* if its transfer function has any of its zeros located outside the unit circle in the z-plane; exponential stability of the system dictates that the poles must be located inside the unit circle. Practical examples of a nonminimum-phase channel include a telephone channel and a fading (multipath) radio channel. In such a situation, channel equalization [i.e., the restoration of the source signal $s(t)$], given only the channel output, is a difficult signal-processing task.

Although the blind signal separation and blind deconvolution problem are usually formulated in their own individual ways, in reality they are types of *inverse problems* with similarities and subtle differences between them. Typically, blind deconvolution involves a single source of information and a single observable (measurement). On the other hand, blind source separation involves multiple but independent sources of information and multiple observables (sensors); preferably, the number of sensors is equal to the number of sources.

1.4 THREE FUNDAMENTAL APPROACHES TO UNSUPERVISED ADAPTIVE FILTERING

Earlier we mentioned that an unsupervised adaptive filter operates by invoking a set of rules that enable the filter to compute an input–output mapping with desirable properties of interest. In this context we can identify three different approaches for the design of unsupervised adaptive filters.

(1) *Bussgang Statistics.* Consider the blind equalization of a communication channel; such a signal-processing operation is a form of blind deconvolution as illustrated in Fig. 1.2. The unsupervised adaptive filter is now configured as shown in Fig. 1.3. The filter has three constituents, a finite-duration impulse-response (FIR) filter, a zero-memory nonlinear output unit, and an algorithm for adjusting the filter's parameters in an iterative manner. The adaptive filtering algorithm is the ubiquitous LMS algorithm described simply as

$$w_k(t+1) = w_k(t) + \mu e(t)x(t-k), \qquad k = -L, \ldots, -1, 0, 1, \ldots, L \tag{1.3}$$

where μ is the step-size (learning-rate) parameter and $e(t)$ is the error

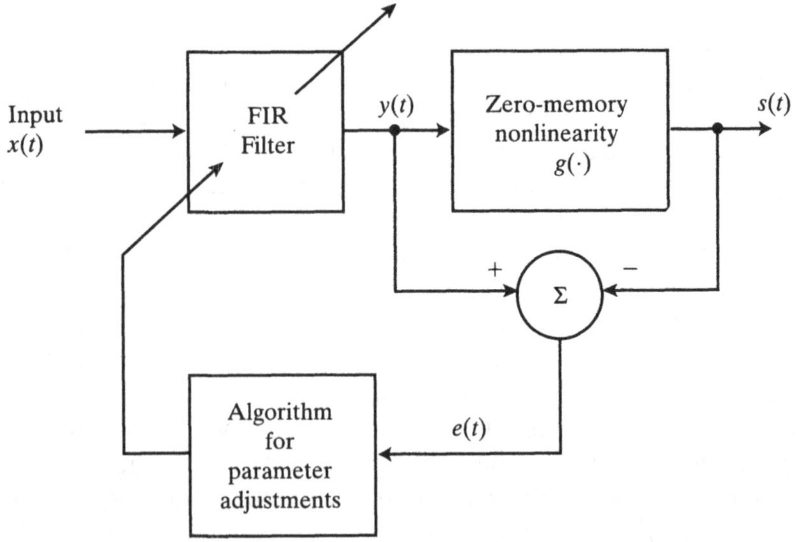

Figure 1.3 Block diagram of a blind adaptive filter of the Bussgang type.

signal defined by

$$e(t) = s(t) - y(t) \tag{1.4}$$

$$= g(y(t)) - y(t) \tag{1.5}$$

and

$$y(t) = \sum_{k=-L}^{L} w_k(t)x(t-k) \tag{1.6}$$

Note that although the LMS algorithm needs a desired response for its operation, the output of the nonlinearity, $s(t)$, is treated as the desired response. Provided that the filter length $(2L+1)$ is sufficiently large and the algorithm has converged, the output of the filter approximately satisfies the following condition:

$$E[y(t)y(t-k)] \simeq E[y(t)g(y(t-k))] \tag{1.7}$$

where the function $g(\cdot)$ is the zero-memory nonlinearity. A process that satisfies this condition is called a *Bussgang process* (Bussgang 1952). In other words, a Bussgang process has the property that its autocorrelation function is equal to the cross-correlation between that process and the output of a zero-memory nonlinearity produced by that process, with both correlations being measured for the same lag. Unsupervised adaptive filters that satisfy Eq. (1.7) are referred to as *Bussgang algorithms.* The Bussgang family of unsupervised adaptive filters include the *decision-directed algorithm* (Lucky 1966), *Sato algorithm* (Sato 1975), and the *constant-modulus algorithm* (CMA) for blind equalization (Godard 1980; Treichler and Agee 1983).

(2) *Higher-Order Statistics.* The Bussgang family of algorithms uses the higher-order statistics of the observed (received) signal in an *implicit* sense. In contrast, the second family of unsupervised adaptive-filtering algorithms uses higher-order statistics of the observed signal in an *explicit* sense. The higher-order statistics of a stationary process are described in terms of *cumulants*, and their Fourier transforms are known as *polyspectra.* Indeed, the cumulants and polyspectra may be viewed as generalizations of the autocorrelation function and power spectrum, respectively. Polyspectra provide a basis for the identification (and therefore blind equalization) of a nonminimum-phase system by virtue of their ability to preserve phase information in the observed signal (Hatzenakos and Nikias 1991).

Let $x(t), x(t+\tau_1), \ldots, x(t+\tau_{l-1})$ denote random variables obtained by observing a stationary stochastic process at times $t, t+\tau_1, \ldots, t+\tau_{l-1}$, respectively. The lth-order cumulant of the process is defined in terms of the joint moments of orders up to l. Specifically, the second-, third-, and fourth-order cumulants are defined by

$$
\begin{aligned}
c_2(\tau) &= E[x(t)x(t+\tau)] \\
c_3(\tau_1, \tau_2) &= E[x(t)x(t+\tau_1)x(t+\tau_2)] \\
c_4(\tau_1, \tau_2, \tau_3) &= E[x(t)x(t+\tau_1)x(t+\tau_2)x(t+\tau_3)] \\
&\quad - E[x(t)x(t+\tau_1)]E[x(t+\tau_2)x(t+\tau_3)] \\
&\quad - E[x(t)x(t+\tau_2)]E[x(t+\tau_1)x(t+\tau_3)] \\
&\quad - E[x(t)x(t+\tau_3)]E[x(t+\tau_1)x(t+\tau_2)]
\end{aligned}
$$

From these definitions, we note the following:

- The second-order cumulant $c_2(\tau)$ is the same as the autocorrelation function of $x(t)$.
- The third-order cumulant $c_3(\tau)$ is the same as the third-order joint moment of $x(t)$.
- The fourth-order cumulant $c_4(\tau_1, \tau_2, \tau_3)$ is different from the fourth-order joint moment of $x(t)$. The difference requires knowledge of six different values of the autocorrelation function of $x(t)$.

Note also that the lth-order cumulant does not depend on time t. For this to be valid, the process $x(t)$ has to be stationary up to order l.

The *lth-order polyspectrum* is defined as the multidimensional Fourier transform of the kth-order cumulant, as shown by

$$
\begin{aligned}
C_k(\omega_1, \omega_2, \ldots, \omega_{l-1}) = \sum_{\tau_1=-\infty}^{\infty} \cdots \sum_{\tau_{l-1}=-\infty}^{\infty} c_k(\tau_1, \tau_2, \ldots, \tau_{l-1}) \\
\cdot \exp(-j(\omega_1\tau_1 + \omega_2\tau_2 + \cdots + \omega_{l-1}\tau_{l-1})) \qquad (1.8)
\end{aligned}
$$

For this polyspectrum to exist we require the lth-order cumulant $c_l(\tau_1, \tau_2, \ldots, \tau_{l-1})$ to be absolutely summable over all $\tau_1, \tau_2, \ldots, \tau_{l-1}$.

Shalvi and Weinstein (1990) have derived a set of criteria based on higher-order cumulants for blind deconvolution. These criteria may be viewed as "universal" in the sense that they do not impose any restrictions on the probability distribution of the source signal, except for the fact that a Gaussian distribution is not permitted. The blind deconvolution algorithms of Shalvi and Weinstein, based on third- and fourth-order cumulants, are of particular interest because of their simplicity.

(3) *Information-Theoretic Models.* The third family of unsupervised adaptive filtering algorithms exploits concepts rooted in Shannon's information theory, namely, entropy and mutual information.

Consider a source of information whose output is denoted by the vector $\mathbf{x}$. Let $f(\mathbf{x})$ denote the probability density of $\mathbf{x}$. The *differential entropy* of this source, signified as X, is defined by

$$\begin{aligned} h(X) &= -E[\log f(\mathbf{x})] \\ &= -\int_{-\infty}^{\infty} f(\mathbf{x}) \log f(\mathbf{x})\, d\mathbf{x} \end{aligned} \tag{1.9}$$

where E is the expectation operator and the integration is carried out over all the components of $\mathbf{x}$. Bell and Sejnowski (1995) have devised a simple and yet effective algorithm for blind source separation, which is based on maximization of the differential entropy defined in a certain way. It is noteworthy that the maximum-entropy algorithm so derived for blind source separation is equivalent to a maximum-likelihood approach (Cardoso 1997).

Mutual information, by definition, involves input and output quantities. Let the vector $\mathbf{y}$ denote the output of a system produced in response to the vector $\mathbf{x}$ applied to the input of the system. The mutual information between these random vectors, denoted by $I(X;Y)$, is defined as the difference between the entropy of $\mathbf{x}$ and the conditional entropy of $\mathbf{x}$ given $\mathbf{y}$, as shown by

$$\begin{aligned} I(X;Y) &= h(X) - h(X|Y) \\ &= -\int_{\infty}^{\infty}\int_{-\infty}^{\infty} f(\mathbf{x},\mathbf{y}) \log\left(\frac{f(\mathbf{x}|\mathbf{y})}{f(\mathbf{x})}\right) d\mathbf{x}\, d\mathbf{y} \end{aligned} \tag{1.10}$$

where $f(\mathbf{x})$ is the probability density function of $\mathbf{x}$; $f(\mathbf{x},\mathbf{y})$ is the joint-probability density function of $\mathbf{x}$ and $\mathbf{y}$; and $f(\mathbf{x}|\mathbf{y})$ is the conditional-probability density function of $\mathbf{x}$ given $\mathbf{y}$. In effect, the entropy $h(X)$ measures the uncertainty that we have about the system input *before* observing the system output, and the conditional entropy $h(X|Y)$ measures our uncertainty about the system input *after* observing the system output. The difference between these two entropies represents the uncertainty about the system input that is resolved by observing the system output.

The mutual information $I(X; Y)$ may also be interpreted as the Kullback-Leibler divergence between two probability distributions, namely, the joint-probability density function $f(\mathbf{x}, \mathbf{y})$ and the product of marginal-probability density functions $f(\mathbf{x})$ and $f(\mathbf{y})$. In general, the *Kullback-Leibler divergence* between two probability density functions $f_1(\mathbf{y})$ and $f_2(\mathbf{y})$ is formally defined by

$$D_{f_1 \| f_2} = \int_{-\infty}^{\infty} f_1(\mathbf{y}) \log\left(\frac{f_1(\mathbf{y})}{f_2(\mathbf{y})}\right) d\mathbf{y} \tag{1.11}$$

The Kullback-Leibler provides a natural basis for solving the blind source-separation problem by viewing $f_1(\mathbf{y})$ as the joint-probability density function of the components constituting the vector $\mathbf{y}$ at the demixer output in Fig. 1.1 and $f_2(\mathbf{y})$ as the product of the marginal probability density functions of those components (Amari et al. 1996; Comon 1994). The Kullback-Leibler divergence between the probability density functions $f_1(\mathbf{y})$ and $f_2(\mathbf{y})$ is also denoted by D_{12}, or $D(f_1, f_2)$.

1.5 ORGANIZATION OF VOLUME II

The main part of Volume II is organized in 3 chapters on blind deconvolution and its relationship to blind source separation.

Chapter 2 by C. R. Johnson, Jr., P. Schniter, I. Fijalkow, L. Tong, J. D. Behm, M. G. Larimore, D. R. Brown, R. A. Casas, T. J. Endres, S. Lambotharan, A. Touzni, H. H. Zeng, M. Green, and J. R. Treichler presents the most detailed treatment of the *constant-modulus algorithm (CMA)* for blind channel equalization available in the literature. Just as the LMS algorithm has established itself as the workhorse for supervised linear adaptive filtering, the CMA has become the workhorse for blind channel equalization. The chapter presents the basic theory of *blind fractionally spaced and/or spatial-diversity equalizers (FSE)* via the CMA. [The idea of fractionally spaced equalization was originally described in the context of equalizers using supervised training (Ungerboeck 1976; Gitlin and Weinstein 1981).] The theory presented in Chapter 2 combines what we already know about the behavior of supervised adaptive equalization and recent results on the robustness of FSE-CMA. Illustrative examples are included in the chapter.

Chapter 3 by Scott Douglas and Simon Haykin explores the relationships between the two related areas: blind deconvolution and blind signal separation. The unifying framework used here is the maximum-likelihood method for parameter estimation. By considering blind signal separation under circulant mixing conditions, relations to blind deconvolution algorithms are established.

Chapter 4 by Constantinos Papadias investigates the necessary and sufficient conditions that have to be satisfied by the observed set of signals in order to guarantee blind source separation. The assumptions made here are that the source outputs are mutually independent, and identically distributed, non-Gaussian, and produced by a linear mixing process. It is shown that the necessary and sufficient conditions for blind source separation in such a scenario depend exclusively on the kurtosis and cross-correlations of the observed signals. Using this set of conditions, criteria for both constrained and unconstrained optimization are described. The algorithms so developed can be viewed as multiuser variants of either the constant modulus algorithm or the kurtosis maximization algorithm due to Shalvi and Weinstein.

REFERENCES

Amari, S., A. Cichocki, and H. H. Yang, 1996, "A new algorithm for blind signal separation," *Adv. Neural Information Processing Systems*, vol. 8, pp. 757–763.

Bell, A. J., and T. J. Sejnowski, 1995, "An information maximization approach to blind separation and blind deconvolution," *Neural Computation*, vol. 6, pp. 1129–1159.

Bussgang, J. J., 1952, *Cross Correlation Functions of Amplitude-Distributed Gaussian Signals*, Technical Report 216, MIT Research Laboratory of Electronics, Cambridge, MA.

Cardoso, J. F., 1997, "Infomax and maximum likelihood for blind source separation," *IEEE Signal Processing Lett.*, vol. 4, pp. 112–114.

Comon, P., 1994, "Independent component analysis: A new concept?" *Signal Processing*, vol. 36, pp. 287–314.

Gitlin, R. D., and S. B. Weinstein, 1981, "Fractionally spaced equalization: an improved digital transversal equalizer," *Bell Syst. Tech. J.*, vol. 60, pp. 275–296.

Godard, D. N., 1980, "Self-recovering equalization and carrier tracking in a two-dimensional data communication system," *IEEE Trans. Communic.*, vol. COM-28, pp. 1867–1875.

Hatzenakos, D., and C. L. Nikias, 1991, "Blind equalization using a trispectrum based algorithm," *IEEE Trans. Communic.*, vol. COM-39, pp. 669–682.

Haykin, S., editor, 1994, *Blind Deconvolution* (Englewood Cliffs; NJ: Prentice-Hall).

Haykin, S., 1996, *Adaptive Filter Theory* (Englewood Cliffs; NJ: Prentice-Hall).

Haykin, S., 1999, *Neural Networks: A Comprehensive Foundation*, Second Edition (Englewood Cliffs; NJ: Prentice-Hall).

Herault, J., C. Jutten, and B. Ans, 1985, "Detection de grandeurs primitives dans un message composite par une architecture de calul neuromimetique un apprentissage non supervise," *Proc. GRETSI*, Nice, France.

Lucky, R. W., 1966, "Techniques for adaptive equalization of digital communication systems," *Bell System Tech. J.*, vol. 45, pp. 255–286.

Sato, Y., 1975, "Two extensional applications of the zero-forcing equalization method," *IEEE Trans. Communic.*, vol. COM-23, pp. 684–687.

Shalvi, O., and E. Weinstein, 1990, "New criteria for blind equalization of nonminimum phase systems (channels)," *IEEE Trans. Inform. Theory*, vol. 36, pp. 312–321.

Treichler, J. R., and B. F. Agee, 1983, "A new approach to multipath correction of constant modulus signals," *IEEE Trans. Acoust. Speech Signal Processing*, vol. ASSP-31, pp. 459–471.

Ungerboeck, G., 1976, "Fractional tap-spacing equalizer and consequences for clock recovery in data modems," *IEEE Trans. Communic.*, vol. COM-24, pp. 856–864.

Widrow, B., and M. C. Hoff, Jr., 1960, "Adaptive switching circuits," *IRE WESCON Convention Record*, Pt. 4, pp. 96–104.

2

THE CORE OF FSE-CMA BEHAVIOR THEORY

C. R. Johnson, Jr., P. Schniter, I. Fijalkow, L. Tong, J. D. Behm, M. G. Larimore, D. R. Brown, R. A. Casas, T. J. Endres, S. Lambotharan, A. Touzni, H. H. Zeng, M. Green, and J. R. Treichler

ABSTRACT

This chapter presents the basics of the current theory regarding the behavior of blind fractionally spaced and/or spatial-diversity equalizers (FSE) adapted via the constant-modulus algorithm (CMA). The CMA, which was developed in the late 1970s and disclosed in the early 1980s, performs a stochastic-gradient descent of a cost function that penalizes the dispersion of the equalizer output from a constant value. The constant modulus (CM) cost function leads to a blind algorithm because evaluation of the CM cost at the receiver does not rely on access to a replica of the transmitted source, as in so-called "trained" scenarios. The capability for blind start-up makes certain communication systems feasible in circumstances that do not admit training. The analytically convenient feature of the fractionally spaced realization of a linear equalizer is the potential for perfect equalization in the absence of channel noise given a finite impulse-response (FIR) equalizer of time span matching that of the FIR channel. The conditions for perfect equalization coupled with some mild conditions on the source can be used to establish con-

Unsupervised Adaptive Filtering, Volume II, Edited by Simon Haykin.
ISBN 0-471-37941-7 © 2000 John Wiley & Sons, Inc.

vergence to perfect performance with FSE parameter adaptation by CMA from any equalizer parameter initialization. The FSE-CMA behavior theory presented here merges the taxonomy of the behavior theory of trained adaptive equalization and recent robustness analysis of FSE-CMA with violation of the conditions leading to perfect equalization and global asymptotic optimality of FSE-CMA.

> *The revolution in data communications technology can be dated from the invention of automatic and adaptive channel equalization in the late 1960s ... Many engineers contributed to this revolution, but the early inventions of Robert W. Lucky, particularly data-driven equalizer adaptation, were the largest factor in realizing higher-speed data communication in commercial equipment.*
>
> —R. D. Gitlin, J. F. Hayes, and S. B. Weinstein, *Data Communication Principles*

> *There are several applications in digital data communications when start-up and retraining of an adaptive equalizer has to be accomplished without the aid of a training sequence. Hence, the system has to be trained "blind" ... We are interested in those circumstances where the eye is closed, and the conventional decision-directed operation will fail ... It is recognized that, in exchange for not requiring data decisions, blind equalization algorithms may require one or two more orders of magnitudes of time to converge. There are two basic algorithms for blind equalization: the constant modulus algorithm (CMA) ...*
>
> —R. D. Gitlin, J. F. Hayes, and S. B. Weinstein, *Data Communication Principles*

2.1 INTRODUCTION

2.1.1 Motivation

The desire to move data at high rates across transmission media with limited bandwidth has prompted the development of sophisticated communications systems, for example, voiceband modems and microwave radio relay systems. Success in those applications has led to great interest in other communication scenarios in which economic or regulatory considerations limit the available transmission bandwidth. An important example of such an application is the wireless and cable distribution of digital television.

Central to the successful employment of most high-data-rate transmission systems is the use of *adaptive equalization* to counteract the disruptive effects of the signal's propagation from the transmitter to the receiver. The equalizer's importance, coupled with the fact that it tends to consume most of the receiver's computational resources and implementation cost, has made it the focus of much analytical and practical attention. Initially, high-data-rate communication systems utilized session-oriented point-to-point links that accommodated cooperative equalizer training. By training, we mean the transmission of a symbol sequence known in advance by the receiver and usually preceded by a clearly identifiable synchronization burst. The more recent emergence of digital multipoint and broadcast systems has produced communication scenarios where training is infeasible or prohibited. In this chapter we are interested in "blind" adaptive equalizers, that is, those that do not need training to achieve convergence from an unacceptable equalizer setting to an acceptable one. In a style intended for the engineer with a first-year-graduate level acquaintance with digital communication systems, this chapter presents the core of the behavior theory of the most popular of blind equalization algorithms, (CMA), in the so-called "fractionally spaced" configuration that dominates current practice. The theory chosen here was selected for its utility in illuminating pragmatic design guidelines.

2.1.2 History

The concept of an adaptive digital linear equalizer was introduced and realized in the 1960s. [See Qureshi (1985) for an excellent survey of pre-1985 advances in trained adaptive equalization and numerous references regarding the highlights cited here.] The received signal's sampling interval matched the baud interval, that is, the time between transmission of consecutive source symbols. The baud-spaced equalizer (BSE) tapped-delay-line length was selected to provide an accurate delayed inverse of a mixed-phase but finite-duration impulse-response (FIR) channel. The common theoretical assumption of infinite equalizer length can be attributed to the recognition that an infinitely long tapped-delay line would be required for perfect[1] equalization of a FIR channel even in the absence of channel noise. Algorithms with and without training were introduced in the 1960s [e.g., Lucky (1966)]. The format with training quickly dominated telephony practice, at least for session start-up, and decision-directed least mean squares (LMS) assumed the role of the fundamental blind method for subsequent tracking.

The 1970s witnessed the emergence of fractionally spaced equalizer

[1] Perfect equalization denotes situations in which the equalizer output sequence equals the transmitted symbol sequence up to a (fixed) unknown amplitude and delay.

(FSE) implementations, that is, those that used sampling rates faster than the source-symbol rate. Improved band-edge equalization capabilities and reduced sensitivity to timing synchronization errors were cited as motivation (Ungerboeck 1976). The practical necessity of "tap leakage" for long FSEs was the most significant adaptive equalizer algorithm modification (Gitlin et al. 1992). Performance analyses for both fractionally spaced and baud-spaced equalizers commonly included assumptions of effectively infinite equalizer length, which permitted perfect equalization and easy translation between time- and frequency-domain interpretations.

During the 1980s, linear equalization methods capable of blind start-up moved from concept into practice. Blind equalization is desirable in multipoint and broadcast systems and necessary in noninvasive test and intercept scenarios. Even in point-to-point communication systems, blind equalization has been adopted for various reasons, including capacity gain and procedural convenience. For performance reasons, fractional spacing of the equalizer became preferred where technologically feasible. However, performance analysis of blind equalizers remained focused almost exclusively on baud-spaced realizations (Haykin 1994).

During the 1990s, blind equalization has been incorporated into several emerging communication technologies (Treichler et al. 1996, 1998), for example, digital cable TV. Also in the 1990s, realization of the ideal capabilities of fractionally-spaced data-adaptive equalizers, especially blind finite-length varieties (Tong et al. 1995), have energized the study of finite-length fractionally spaced blind equalizers [See, for example, Johnson et al. (1998), appearing in the special issue (Liu 1998)]. The advantages that result from utilizing time-diversity systems (i.e., fractional sampling) also occur in spatial-diversity systems (e.g., those employing multiple sensors or cross polarity) and code-diversity systems (e.g., short-code DS-CDMA) (Paulraj and Papadias 1997).

2.1.3 Our Goal: Behavior Theory Basics and Design Guidelines

The pedagogy employed here is fundamentally similar to that used in a variety of widely cited textbooks (e.g., Proakis 1995; Gitlin et al. 1992; Lee and Messerschmitt 1994) for trained, baud-spaced equalization theory, based on minimization of the mean squared error (MSE) in recovery of the training sequence. This approach is bolstered by recent results—to be described in this chapter—on the similarity of the locations of MSE minima and minima of the (blind) constant-modulus (CM) cost function. Such similarities prompt the adoption of a taxonomy associated with design rules for trained stochastic-gradient descent procedures, such as the LMS algorithm, to a stochastic-gradient descent approach for minimizing the CM cost via the CMA. This results in design guidelines—to be developed and dissected in this chapter—regarding adaptive algorithm step-size selection, equalizer length, and equalizer parameter (re)initialization.

2.1.4 Content Map

Against this backdrop we present a map of the contents of this chapter, carrying us from an FSE problem formulation to an understanding of the design guidelines for blind CMA-FSE, that is, CMA-based adaptation of an FSE.

- Section 2.2 introduces the FSE problem formulation. A multichannel model is adopted and the capability for perfect symbol recovery is established with an FSE in the absence of channel noise. With the addition of channel noise, the Wiener solution (with a necessarily nonzero minimum mean-squared, delayed-source recovery error) is formulated for an infinite-duration impulse-response (IIR) linear equalizer and for a FIR linear equalizer. The resulting minimum MSE is dissected in terms of its factors (i.e., noise power, equalizer length, channel convolution matrix singular values, and target system delay). Given a description of transient and asymptotic performance of the underlying average system behavior, a brief distillation of step-size and equalizer-length design guidelines for LMS-FSE is also provided as the background against which CMA-FSE design guidelines will be composed.
- Section 2.3 begins with a definition of the CM (or CMA 2-2) criterion, and combines the perfect equalization requirements with some generic assumptions on the source statistics to result in a set of requirements for CMA-FSE's global asymptotic optimality. A series of 2-tap FSE CM cost functions and CMA trajectories is used to illustrate the basic robustness properties with violations of each of these conditions. Approximate perturbation analyses of the effects of channel noise and equalizer length and a geometric analysis of the achieved MSE of a CM-minimizing equalizer in the presence of channel noise are exploited for insight. Differences in CMA-FSE relative to LMS-FSE are highlighted with examination of convergence rate and excess MSE.
- Section 2.4 focuses on three design choices in CMA-FSE implementation: adaptive equalizer parameter update step-size, equalizer

length, and equalizer parameter initialization. Guidelines are developed through an example-driven tutorial approach. Single-spike initialization for CMA-BSE and double-spike initialization for CMA-FSE are discussed in terms of magnitude and location. In step-size selection, the trade-offs in LMS design (i.e., (1) between convergence rate and excess MSE, and (2) between tracking error and gradient approximation error) are noted to drive CMA step-size selection as well. A similar trade-off between improved modeling accuracy and increased excess mean squared error is discussed for increases in equalizer length.

- Section 2.5 presents three case studies, each of which yields a blind equalizer capable of dealing with a particular problem class (specifically, voice-channel modem, cable-borne HDTV, and microwave radio) represented by signals and channel models in a publicly accessible database.

2.1.5 Notation

This subsection contains the abbreviations and mathematical notation used throughout this chapter.

ACRONYMS AND ABBREVIATIONS

Acronym	Definition
BER	Bit error rate
BPSK	Binary phase-shift keying
BS	Baud-spaced
BSE	Baud-spaced equalizer
CM	Constant modulus
CMA	Constant modulus algorithm
DD	Decision-directed
EMSE	Excess mean-squared error
FIR	Finite-duration impulse response
FS	Fractionally spaced
FSE	Fractionally spaced equalizer
i.i.d.	Independent and identically distributed
IIR	Infinite-duration impulse response
ISI	Intersymbol interference
LMS	Least mean square
MMSE	Minimum mean-squared error
MSE	Mean-squared error
ODE	Ordinary differential equation

Acronym	Definition
PAM	Pulse-amplitude modulation
PBE	Perfect blind equalizability
pdf	Probability density function
PSK	Phase-shift keying
QAM	Quadrature amplitude modulation
QPSK	Quadrature phase-shift keying
SER	Symbol error rate
SNR	Signal-to-noise ratio
SPIB	Signal-Processing Information Base
SVD	Singular value decomposition
ZF	Zero forcing

MATHEMATICAL NOTATION

Symbol	Definition
$(\cdot)^T$	Transposition
$(\cdot)^*$	Conjugation
$(\cdot)^H$	Hermitian transpose (i.e., conjugate transpose)
$(\cdot)^\dagger$	Moore-Penrose pseudoinverse
$\mathrm{tr}(\cdot)$	Trace operator
$\lambda_{\min}(\cdot)$	Minimum eigenvalue
$\lambda_{\max}(\cdot)$	Maximum eigenvalue
$\Vert\mathbf{x}\Vert_p$	ℓ_p norm: $\sqrt[p]{\sum_n \lvert x_n\rvert^p}$
$\Vert\mathbf{x}\Vert_{\mathbf{A}}$	Norm defined by $\sqrt{\mathbf{x}^H\mathbf{A}\mathbf{x}}$ for positive-definite Hermitian $\mathbf{A}$
$\Vert\mathbf{x}(e^{j\omega})\Vert$	L_2 norm: $\sqrt{\sum_n \lvert x_n(e^{j\omega})\rvert^2}$
$\mathbf{I}$	Identity matrix
$\mathbf{e}_i$	Column vector with 1 at the ith entry ($i \geq 0$) and zeros elsewhere
$\mathbb{R}$	The field of real-valued scalars
$\mathbb{C}$	The field of complex-valued scalars
$\mathrm{Re}\{\cdot\}$	Extraction of real-valued component
$\mathrm{Im}\{\cdot\}$	Extraction of imaginary-valued component
$Z\{\cdot\}$	z-transform, i.e., $Z\{x_n\} = \sum_n x_n z^{-n}$ for allowable $z \in \mathbb{C}$
$\mathrm{E}\{\cdot\}$	Expectation
$\nabla_{\mathbf{f}}$	Gradient with respect to $\mathbf{f}$: $\nabla_{\mathbf{f}} = \dfrac{\partial}{\partial \mathbf{f}_r} + j\dfrac{\partial}{\partial \mathbf{f}_i}$ where $\mathbf{f}_r = \mathrm{Re}\{\mathbf{f}\}$ and $\mathbf{f}_i = \mathrm{Im}\{\mathbf{f}\}$
$\mathscr{H}_{\mathbf{f}}$	Hessian with respect to $\mathbf{f}$: $\mathscr{H}_{\mathbf{f}} = \nabla_{\mathbf{f}}\nabla_{\mathbf{f}^T}$, for real-valued $\mathbf{f}$
	Boldface lowercase Roman typeface designates vectors
	Boldface uppercase Roman typeface designates matrices

SYSTEM MODEL QUANTITIES

Symbol	Definition
T	Symbol period
n	Index for quantities sampled at baud intervals: $t = nT$
k	Index for fractionally sampled quantities: $t = kT/P$
δ	System delay (nonnegative, integer-valued)
q_n	System impulse response coefficient
s_n	Source symbol
y_n	System/equalizer output
v_n	Filtered noise contribution to system output
$q(z)$	System transfer function $Z\{q_n\}$
$s(z)$	z-transformed source sequence $Z\{s_n\}$
$y(z)$	z-transformed output sequence $Z\{y_n\}$
$v(z)$	z-transformed noise sequence $Z\{v_n\}$
$\mathbf{q}$	Vector of BS system response coefficients $\{q_n\}$
$\mathbf{s}(n)$	Vector of past source symbols $\{s_n\}$ at time n
$\mathbf{H}$	(Multi)channel convolution matrix
σ_s^2	Variance of source sequence: $\mathbf{E}\{\lvert s_n\rvert^2\}$
κ_s	Normalized kurtosis of source process: $\mathbf{E}\{\lvert s_n\rvert^4\}/\sigma_s^4$
κ_g	Normalized kurtosis of a Gaussian process

MULTIRATE MODEL QUANTITIES

Symbol	Definition
P	Fractional sampling factor
h_k	Channel impulse-response coefficient
f_k	Equalizer impulse-response coefficient
w_k	Additive channel noise sample
r_k	Channel output (i.e., receiver input) sample
x_k	Noiseless channel output sample
$\mathbf{h}$	Vector of FS channel coefficients $\{h_k\}$
$\mathbf{f}$	Vector of FS equalizer coefficients $\{f_k\}$
$\mathbf{w}(n)$	Vector of FS channel noise samples $\{w_k\}$ at time n
$\mathbf{r}(n)$	Vector of FS channel outputs $\{r_k\}$ at time n
$\mathbf{x}(n)$	Vector of noiseless FS channel outputs $\{x_k\}$ at time n
σ_w^2	Variance of additive noise process: $\mathrm{E}\{\lvert w_k\rvert^2\}$
σ_r^2	Variance of FS received signal: $\mathrm{E}\{\lvert r_k\rvert^2\}$
κ_w	Normalized kurtosis of additive noise process: $\mathrm{E}\{\lvert w_k\rvert^4\}/\sigma_w^4$

MULTICHANNEL MODEL QUANTITIES

Symbol	Definition
P	Number of subchannels
L_h	Order of subchannel polynomials
L_f	Order of subequalizer polynomials
L_g	Order of subchannel polynomials' GCD
$h_n^{(p)}$	Impulse-response coefficient n of pth subchannel
$f_n^{(p)}$	Impulse-response coefficient n of pth subequalizer
$w_n^{(p)}$	Noise sample added to pth subchannel at time n
$r_n^{(p)}$	Output of subchannel p at time n
$x_n^{(p)}$	Noiseless output of subchannel p at time n
$\mathbf{h}^{(p)}$	Vector of BS subchannel response coefficients $\{h_n^{(p)}\}$
$\mathbf{f}^{(p)}$	Vector of BS subequalizer response coefficients $\{f_n^{(p)}\}$
$h^{(p)}(z)$	Transfer function of pth subchannel $Z\{h_n^{(p)}\}$
$f^{(p)}(z)$	Transfer function of pth subequalizer $Z\{f_n^{(p)}\}$
$\mathbf{h}_n$	Vector-valued channel impulse-response coefficient
$\mathbf{f}_n$	Vector-valued equalizer impulse-response coefficient
$\mathbf{w}_n$	Vector-valued additive channel noise sample
$\mathbf{r}_n$	Vector-valued channel output (i.e., receiver input) sample
$\mathbf{x}_n$	Vector-valued noiseless channel output sample
$\mathbf{h}(z)$	Vector-valued channel transfer function $Z\{\mathbf{h}_n\}$
$\mathbf{f}(z)$	Vector-valued equalizer transfer function $Z\{\mathbf{f}_n\}$
$\mathbf{w}(z)$	Vector-valued z-transform of noise $Z\{\mathbf{w}_n\}$
$\mathbf{r}(z)$	Vector-valued z-transform of channel output $Z\{\mathbf{r}_n\}$
$\mathbf{x}(z)$	Vector-valued z-transform of noiseless channel output $Z\{\mathbf{x}_n\}$

EQUALIZER DESIGN QUANTITIES

Symbol	Definition
$J_{\mathrm{m}}^{(\delta)}(\cdot)$	MSE cost function for system delay δ
$J_{\mathrm{cm}}(\cdot)$	CM cost function
$\mathscr{E}_{\mathrm{m}}^{(\delta)}$	MMSE associated with system delay δ
$\mathscr{E}_{\chi}$	Excess MSE
$\mathbf{f}_{\mathrm{z}}^{(\delta)}$	Zero-forcing equalizer associated with system delay δ
$\mathbf{f}_{\mathrm{m}}^{(\delta)}$	Wiener equalizer associated with system delay δ
$\mathbf{f}_{\mathrm{c}}^{(\delta)}$	CM equalizer associated with system delay δ
$\mathbf{q}_{\mathrm{z}}^{(\delta)}$	System response achieved by ZF equalizer $\mathbf{f}_{\mathrm{z}}^{(\delta)}$
$\mathbf{q}_{\mathrm{m}}^{(\delta)}$	System response achieved by Wiener equalizer $\mathbf{f}_{\mathrm{m}}^{(\delta)}$

$\mathbf{q}_c^{(\delta)}$	System response achieved by CM equalizer $\mathbf{f}_c^{(\delta)}$
$\mathbf{R}_{r,r}$	Received signal autocorrelation matrix: $E\{\mathbf{r}(n)\mathbf{r}^H(n)\}$
$\mathbf{R}_{x,x}$	Noiseless received signal autocorrelation matrix: $E\{\mathbf{x}(n)\mathbf{x}^H(n)\}$
$\mathbf{d}_{r,s}^{(\delta)}$	Cross-correlation between the received and desired signal: $E\{\mathbf{r}(n)s_{n-\delta}\}$
τ_{cma}	Time constant of CMA local convergence
μ	Step size used in LMS and CMA
γ	CMA dispersion constant

2.2 MMSE EQUALIZATION AND LMS

This section formulates the communications channel model and the FSE problem. In addition, it highlights basic results for zero-forcing and minimum mean-square-error (MMSE) linear equalizers and their adaptive implementation via the LMS stochastic-gradient descent algorithm. Section 2.3 will leverage these concepts to draw a parallel with CM receiver theory.

Although MMSE equalization is, in general, not optimal in the sense of minimizing symbol error rate (SER), it is perhaps the most widely used method in modem (among various other communication system) designs for intersymbol interference (ISI) limited channels. Theoretically, the combination of coding and linear MMSE equalization offers a practical way to achieve channel capacity (even when SER is not minimized!) (Cioffi et al. 1995). One advantage of the mean-squared-error (MSE) cost function is that it is quadratic and therefore unimodal (i.e., it is not complicated by the possibility of an MSE-minimizing algorithm (e.g., LMS) converging to a false local minimum). With all the merits of MMSE equalization, we will be motivated to compare blind equalizers, such as those minimizing the CM criterion, to MMSE equalizers.

2.2.1 Channel Models

Consider the equalization of linear, time-invariant, FIR channels transmitting an information symbol sequence $\{s_n\}$ as shown in Fig 2.1a. In a

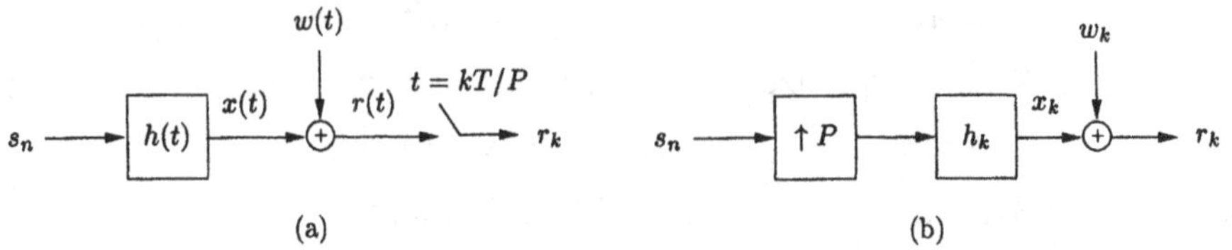

Figure 2.1 Two equivalent single-sensor models: (*a*) the continuous-time channel model. and (*b*) the discrete-time multirate channel model.

single-sensor scenario, the continuous-time received baseband signal $r(t)$ has the following form:

$$r(t) = \sum_{i=-\infty}^{\infty} s_i h(t - iT) + w(t) \tag{2.1}$$

where T is the symbol period, $h(t)$ is the continuous-time channel impulse response, and $w(t)$ represents additive channel noise. For simplicity, $w(t)$ is typically assumed to be a white Gaussian noise process. The model of the channel impulse response includes the (possibly unknown) pulse-shaping filter at the transmitter, the impulse response of the linear approximation to the propagation channel, and the receiver front-end filter (i.e., any filter prior to the equalizer).

The discrete-time *multirate* channel model shown in Fig. 2.1*b* is obtained by uniformly sampling $r(t)$ at an integer[2] fraction of the symbol period, T/P. The fractionally spaced (FS) channel output is then given by

$$r_k \triangleq r\left(k\frac{T}{P}\right) = \sum_i s_i \underbrace{h\left(k\frac{T}{P} - iT\right)}_{h_{k-iP}} + \underbrace{w\left(k\frac{T}{P}\right)}_{w_k} \tag{2.2}$$

$$= \underbrace{\sum_i s_i\, h_{k-iP}}_{x_k} + w_k \tag{2.3}$$

where the x_k are the (FS) noiseless channel outputs, the h_k are FS samples of the channel impulse response, and the w_k are FS samples of the channel noise process. [Throughout the chapter, the index n is reserved for baud-spaced (BS) quantities, while the index k is applied to fractionally spaced quantities.] For finite-duration channels, it is convenient to collect the fractionally sampled channel response coefficients into the vector

$$\mathbf{h} = (h_0, h_1, h_2, \ldots, h_{(L_h+1)P-1})^T \tag{2.4}$$

where L_h denotes the length of the channel impulse response in symbol intervals.

A particularly useful equivalent to the multirate model is the symbol-rate *multichannel* model shown in Fig. 2.2, where the pth subchannel $(p = \{1, \ldots, P\})$ is obtained by subsampling $\mathbf{h}$ by the factor P. The re-

[2] In general, fractionally spaced equalizers may operate at noninteger multiples of the baud rate. For simplicity, however, we restrict our attention to integer multiples.

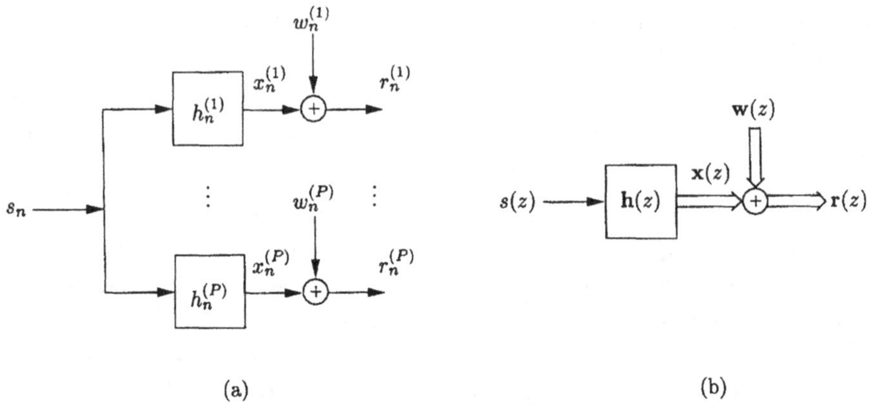

Figure 2.2 (*a*) Multichannel model, and (*b*) its *z*-domain equivalent.

spective multichannel quantities are, for the pth subchannel,

$$\begin{aligned} h_n^{(p)} &\triangleq h_{(n+1)P-p}, & x_n^{(p)} &\triangleq x_{(n+1)P-p}, \\ r_n^{(p)} &\triangleq r_{(n+1)P-p}, & w_n^{(p)} &\triangleq w_{(n+1)P-p} \end{aligned} \tag{2.5}$$

Denoting the vector-valued channel response samples (at baud index n) and their z-transform by

$$\mathbf{h}_n \triangleq \begin{pmatrix} h_n^{(1)} \\ \vdots \\ h_n^{(P)} \end{pmatrix} \xrightarrow{Z} \mathbf{h}(z) \tag{2.6}$$

we arrive at the following system of equations (in both time- and z-domains):

$$\mathbf{x}_n = \sum_{i=0}^{L_h} \mathbf{h}_i s_{n-i} \qquad \mathbf{r}_n = \mathbf{x}_n + \mathbf{w}_n \tag{2.7}$$

$$\mathbf{x}(z) = \mathbf{h}(z)s(z) \qquad \mathbf{r}(z) = \mathbf{x}(z) + \mathbf{w}(z) \tag{2.8}$$

Here, $\mathbf{x}_n$ denotes the vector-valued multichannel output without noise, while $\mathbf{r}_n$ denotes the (noisy) received vector signal. Note that the multichannel vector quantities are indexed at the baud rate. We shall find this multichannel structure convenient in the sequel.

Though our derivation of the multichannel model originates from the single-sensor application of Fig. 2.1, the multichannel formulation applies directly to situations in which multiple sensors are used, with or without oversampling [see, e.g., Moulines et al. (1995)]. In other words,

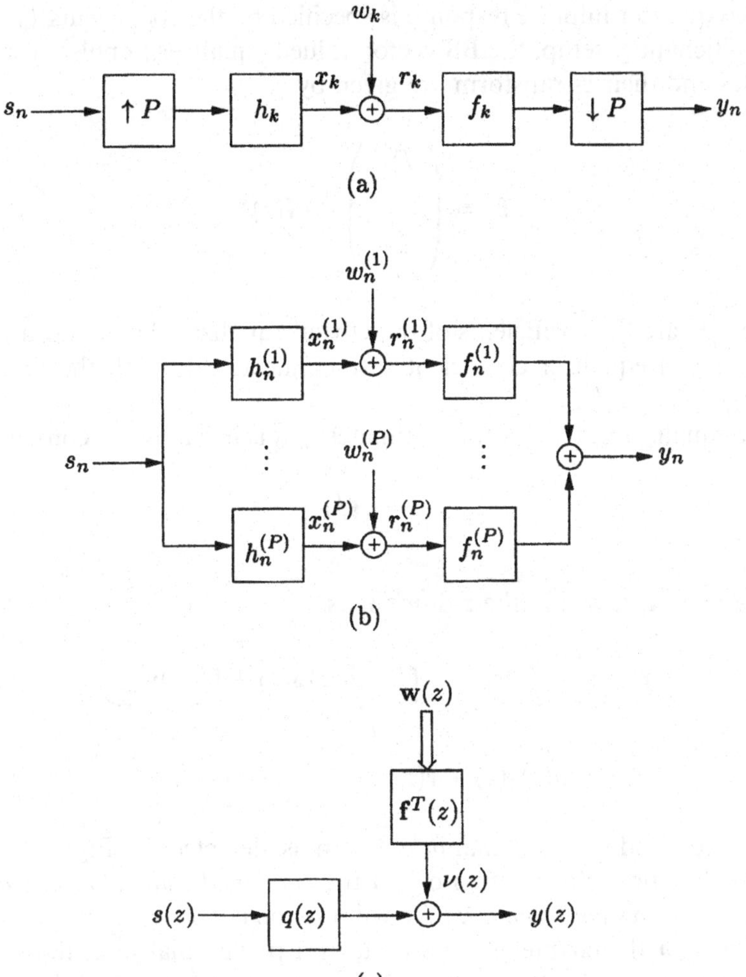

Figure 2.3 Equivalent system models: (*a*) multirate, (*b*) multichannel, and (*c*) their *z*-domain representation.

the disparity achieved by oversampling in time or in space results in the same mathematical model. This is evident in Fig. 2.2, where $h_n^{(p)}$ would characterize the impulse-response coefficients of the (BS) physical channel from the transmitter to the pth sensor.

2.2.2 Linear Equalizers

For a fixed system delay δ, an FSE $\mathbf{f}$ is a linear estimator of the input $s_{n-\delta}$ given the multirate observation r_k or, equivalently, the multichannel observation $\mathbf{r}_n$ (both shown in Fig. 2.3). In the multirate setup, the T/P-

spaced equalizer impulse response is specified by the coefficients $\{f_k\}$. In the multichannel setup, the BS vector-valued equalizer impulse-response samples and their z-transform are given by

$$\mathbf{f}_n \triangleq \begin{pmatrix} f_n^{(1)} \\ \vdots \\ f_n^{(P)} \end{pmatrix} \xrightarrow{Z} \mathbf{f}(z) \tag{2.9}$$

where $f_n^{(p)}$ are the coefficients of the pth subequalizer. The multirate and multichannel equalizer coefficients are connected through the relationship $f_{nP+p-1} = f_n^{(p)}$.

The equalizer output y_n that estimates $s_{n-\delta}$ is given by the convolution

$$y_n = \sum_i \mathbf{f}_i^T \mathbf{r}_{n-i} \tag{2.10}$$

and can be expressed in the z-domain as

$$y(z) = \mathbf{f}^T(z)\mathbf{r}(z) = \underbrace{\mathbf{f}^T(z)\mathbf{h}(z)}_{q(z)} s(z) + \underbrace{\mathbf{f}^T(z)\mathbf{w}(z)}_{v(z)} \tag{2.11}$$

$$= q(z)s(z) + v(z) \tag{2.12}$$

The corresponding multichannel system is depicted in Fig. 2.3. The transfer function $q(z)$ is often called the *combined channel-equalizer response* or the *system response*. (We will use the latter terminology for the remainder of the chapter.) Note that, as a polynomial in z, the system response is a baud-rate quantity. It is important to realize that, once restrictions are placed on the channel and/or equalizer, not all system responses may be attainable.

Next we consider the case where the channel and equalizer impulse responses are restricted to be finite in duration. In such a case, the estimate of $s_{n-\delta}$ is obtained from the past $L_f + 1$ multichannel observations $\mathbf{r}_n$, where $L_f + 1$ denotes the length of the multichannel equalizer. Specifically, we have

$$y_n = \sum_{i=0}^{L_f} \mathbf{f}_i^T \mathbf{r}_{n-i} = \mathbf{f}^T \mathbf{r}(n) \tag{2.13}$$

where

$$\mathbf{f} \triangleq \begin{pmatrix} \mathbf{f}_0 \\ \vdots \\ \mathbf{f}_{L_f} \end{pmatrix} = \begin{pmatrix} f_0 \\ \vdots \\ f_{P(L_f+1)-1} \end{pmatrix} \quad \text{and}$$

$$\mathbf{r}(n) \triangleq \begin{pmatrix} \mathbf{r}_n \\ \vdots \\ \mathbf{r}_{n-L_f} \end{pmatrix} = \begin{pmatrix} r_{(n+1)P-1} \\ \vdots \\ r_{(n-L_f)P} \end{pmatrix} \tag{2.14}$$

The vector-valued received signal in Eq. (2.7) can be written as

$$\underbrace{\begin{pmatrix} \mathbf{r}_n \\ \vdots \\ \mathbf{r}_{n-L_f} \end{pmatrix}}_{\mathbf{r}(n)} = \underbrace{\begin{pmatrix} \mathbf{h}_0 & \cdots & \mathbf{h}_{L_h} & & \\ & \ddots & & \ddots & \\ & & \mathbf{h}_0 & \cdots & \mathbf{h}_{L_h} \end{pmatrix}}_{\mathbf{H}} \underbrace{\begin{pmatrix} s_n \\ \vdots \\ s_{n-L_f-L_h} \end{pmatrix}}_{\mathbf{s}(n)} + \underbrace{\begin{pmatrix} \mathbf{w}_n \\ \vdots \\ \mathbf{w}_{n-L_f} \end{pmatrix}}_{\mathbf{w}(n)} \tag{2.15}$$

$$\mathbf{r}(n) = \mathbf{H}\mathbf{s}(n) + \mathbf{w}(n) \tag{2.16}$$

where $\mathbf{H}$ is often referred to as the channel matrix. As evident in Eq. (2.14), our construction ensures that the fractionally sampled coefficients of the vector quantities $\mathbf{r}(n)$, $\mathbf{w}(n)$, $\mathbf{x}(n) \triangleq \mathbf{H}\mathbf{s}(n)$, and $\mathbf{f}$ are well-ordered with respect to the multirate time-index.[3] Substituting the received vector expression, Eq. (2.16), into Eq. (2.13), we obtain the following system output, occurring at the baud rate:

$$y_n = \mathbf{f}^T\mathbf{H}\mathbf{s}(n) + \mathbf{f}^T\mathbf{w}(n) \tag{2.17}$$

$$= \mathbf{q}^T\mathbf{s}(n) + \nu_n \tag{2.18}$$

The vector $\mathbf{q}$ represents the system impulse response (whose coefficients are sampled at the baud rate), and the quantity ν_n denotes the filtered channel-noise contribution to the system output.

In measuring the performance of linear equalizers, we will be considering the MSE criterion. Given a fixed system delay δ for mutually uncorrelated symbol and noise processes with variances $\mathrm{E}\{|s_n|^2\} = \sigma_s^2$ and

[3] Note that the ordering of $\mathbf{h}_n$ in Eq. (2.6) implies $\mathbf{h} = (h_0, h_1, \ldots, h_{(L_h+1)P-1})^T \neq (\mathbf{h}_0^T, \mathbf{h}_1^T, \ldots, \mathbf{h}_{L_h}^T)^T$.

$E\{|w_k|^2\} = \sigma_w^2$, the MSE achieved by linear equalizer **f** is defined by

$$J_{\mathrm{m}}^{(\delta)}(\mathbf{f}) \triangleq E\{|\mathbf{f}^T\mathbf{r}(n) - s_{n-\delta}|^2\} \tag{2.19}$$

$$= E\{|\mathbf{q}^T\mathbf{s}(n) - s_{n-\delta} + \mathbf{f}^T\mathbf{w}(n)|^2\} \tag{2.20}$$

$$= \underbrace{\sigma_s^2\|\mathbf{q} - \mathbf{e}_\delta\|_2^2}_{\text{ISI and bias}} + \underbrace{\sigma_w^2\|\mathbf{f}\|_2^2}_{\text{noise enhancement}} \tag{2.21}$$

where $\mathbf{e}_\delta$ denotes a vector with a one in the δth position ($\delta \geq 0$) and zeros elsewhere.

From Eq. (2.21) we note that the MSE of a linear equalizer comes from two sources: (1) ISI plus bias, and (2) noise enhancement. ISI measures the effect of residual interference from other transmitted symbols, while bias refers to an incorrect amplitude estimation of the desired symbol. Both are minimized by making the system response close to the unit impulse, which leads to the so-called *zero-forcing equalizer* [see, e.g., Lee and Messerschmitt (1994)]. For certain channels, however, reducing ISI and bias leads to an increase in equalizer norm, and thus an enhancement in noise power. The *MMSE equalizer* achieves the optimal trade-off between ISI reduction and noise enhancement (for a particular noise level) in the sense of minimizing $J_m^{(\delta)}(\mathbf{f})$. In Section 2.3, we shall discuss how CM receivers, not designed to minimize the MSE criterion, achieve a similar compromise. In the remainder of the chapter, the terms "receiver" and "equalizer" will be used interchangeably.

2.2.3 Zero-Forcing Receivers

An equalizer capable of *perfect symbol recovery* in the absence of noise, that is, $y_n = s_{n-\delta}$ for some fixed delay δ, is called a zero-forcing (ZF) equalizer (Lucky 1966) and is denoted by $\mathbf{f}_{\mathrm{z}}^{(\delta)}$. From Eq. (2.12), we see that perfect symbol recovery requires $v(z) = 0$ and $q(z) = z^{-\delta}$, implying an absence of channel noise and a particular relationship between the subchannels and subequalizers (discussed in detail below).

For nontrivial channels with finite-duration impulse response, BS ZF equalizers of finite-duration impulse response do not exist for reasons that will become evident in the discussion below. In contrast, FIR fractionally spaced ZF equalizers do exist under particular conditions. A sufficient condition for perfect symbol recovery, referred to as *strong perfect equalization* (SPE), is that *any* system response **q** can be achieved through proper choice of equalizer **f**. Applicable to both FIR and

IIR channels, SPE guarantees perfect symbol recovery for any delay $0 \leq \delta \leq L_f + L_h$ (in the absence of noise). The SPE requirement may seem overly strong since it may be satisfactory to attain perfect equalization at only one particular delay. However, the class of channels allowing perfect symbol recovery for a restricted range of delays consists of primarily trivial channels. In other words, without SPE, it is usually impossible to achieve perfect symbol recovery for a fixed delay.

A necessary and sufficient condition for SPE is that the channel matrix **H** is full column rank, which has implications for the subchannel and subequalizer polynomials $h^{(p)}(z) = Z\{h_n^{(p)}\}$ and $f^{(p)}(z) = Z\{f_n^{(p)}\}$, respectively. A fundamental requirement for SPE is that the subchannel polynomials must not all share a common zero, that is, $\{h^{(p)}(z)\}$ must be coprime.[4] This is often described by the condition: $\forall z$, $\mathbf{h}(z) \neq 0$. It can be shown (Tong et al. 1995) that when the $\{h^{(p)}(z)\}$ are coprime, there exists a minimum equalizer length for which the channel matrix **H** has full column rank, thus ensuring SPE. Specifically, $L_f \geq L_h - 1$ is a sufficient equalizer length condition when the subchannels are coprime.

The subchannel polynomials are coprime if and only if there exists a set $\{f^{(p)}(z)\}$ that satisfies the Bezout equation (Fuhrmann 1996; Kailath 1980):

$$1 = \sum_{p=1}^{P} f^{(p)}(z)h^{(p)}(z) = \mathbf{f}^T(z)\mathbf{h}(z) \tag{2.22}$$

In other words, equalizer polynomials that satisfy the Bezout equation specify ZF equalizers.

We summarize our statements about perfect equalization in the following set of equivalences, valid in the absence of channel noise:

- Satisfaction of SPE conditions;
- Channel matrix **H** of full column rank;
- Existence of ZF equalizer for all system delays δ, where $0 \leq \delta \leq L_f + L_h$.
- Bezout equation satisfied.

To gain further insight into the SPE condition, it is useful to examine what happens when the subchannels are not coprime. For example,

[4] Note that while coprimeness ensures the absence of any zero common to *all* subchannels in the set $\{h^{(p)}(z)\}$, it allows the existence of zeros common to strict subsets of $\{h^{(p)}(z)\}$.

consider the case where $g(z) = 1 + g_1 z^{-1}$ can be factored out of every subchannel polynomial $\{h^{(1)}(z), \ldots, h^{(P)}(z)\}$, leaving $\{\bar{h}^{(1)}(z), \ldots, \bar{h}^{(P)}(z)\}$. It becomes clear that, for any set $\{f^{(p)}(z)\}$,

$$\sum_{p=1}^{P} f^{(p)}(z) h^{(p)}(z) = (1 + g_1 z^{-1}) \sum_{p=1}^{P} f^{(p)}(z) \bar{h}^{(p)}(z) \neq 1 \tag{2.23}$$

Thus, the presence of the common subchannel factor $g(z)$ prevents the Bezout equation from being satisfied, making the ideal system response $q(z) = 1$ unattainable.

When the subchannels are not coprime, it may still be possible to approximate the perfect system response with a finite-length equalizer. In this case the equalizer is designed so that the remaining subchannel–subequalizer combinations approximate a delayed inverse of $g(z)$, that is,

$$\sum_{p=1}^{P} f^{(p)}(z) \bar{h}^{(p)}(z) \approx g^{-1}(z) z^{-\delta} \tag{2.24}$$

In general, the approximation improves as the equalizer length is increased, though performance will depend on the choice of system delay $\delta : 0 \leq \delta \leq L_f + L_{\bar{h}}$. The implication here is that long-enough equalizers can well approximate ZF equalizers even in the presence of common subchannel roots (as long as the common roots do not lie on the z-plane's unit circle).

We can also examine the effect of common zero(s) in the time domain via a decomposition of the channel matrix $\mathbf{H}$. If an order L_g polynomial $g(z)$ can be factored out of every subchannel, then it can be factored out of each row of the vector polynomial $\mathbf{h}(z)$, leaving $\bar{\mathbf{h}}(z)$ (of order $L_h - L_g$). We exploit this in the decomposition $\mathbf{H} = \bar{\mathbf{H}}\mathbf{G}$, where

$$\bar{\mathbf{H}} = \begin{pmatrix} \bar{\mathbf{h}}_0 & \cdots & \bar{\mathbf{h}}_{L_h - L_g} & & \\ & \ddots & & \ddots & \\ & & \bar{\mathbf{h}}_0 & \cdots & \bar{\mathbf{h}}_{L_h - L_g} \end{pmatrix} \quad \text{and}$$
$$\mathbf{G} = \begin{pmatrix} g_0 & \cdots & g_{L_g} & & \\ & \ddots & & \ddots & \\ & & g_0 & \cdots & g_{L_g} \end{pmatrix} \tag{2.25}$$

The matrix $\bar{\mathbf{H}}$ is full column rank with dimension $P(L_f + 1) \times$

$(L_f + L_h - L_g + 1)$, while $\mathbf{G}$ is full row rank with dimension $(L_f + L_h - L_g + 1) \times (L_f + L_h + 1)$. Since the rank of $\overline{\mathbf{H}}$ cannot exceed $(L_f + L_h - L_g + 1)$ and $\mathbf{H}$ has $(L_f + L_h + 1)$ columns, the choice of $L_g > 0$ prevents $\mathbf{H}$ from achieving full column rank.

Finally, it is worth mentioning that the presence of a common subchannel root is associated with P roots of the FS channel polynomial (i.e., $h_0 + h_1 z^{-1} + \cdots + h_{P(L_h+1)} z^{-P(L_h+1)}$) lying equally spaced on a circle in the complex plane (Tong et al. 1995). In the case of $P = 2$, this implies that common subchannel roots are equivalent to FS channel roots reflected across the origin.

2.2.4 Wiener Receivers

For a fixed system delay δ, the Wiener receiver $\mathbf{f}_{\mathrm{m}}^{(\delta)}$ estimates the source symbol $s_{n-\delta}$ by minimizing the MSE cost

$$J_{\mathrm{m}}^{(\delta)}(\mathbf{f}) = \mathrm{E}\{|\mathbf{f}^T \mathbf{r}(n) - s_{n-\delta}|^2\} \tag{2.26}$$

$$\mathbf{f}_{\mathrm{m}}^{(\delta)} \triangleq \arg \min_{\mathbf{f}} J_{\mathrm{m}}^{(\delta)}(\mathbf{f}) \tag{2.27}$$

For notational simplicity, the remainder of Section 2.2 assumes that the input s_n is a zero-mean, unit-variance ($\sigma_s^2 = \mathrm{E}\{|s_n|^2\} = 1$), uncorrelated random process. Furthermore, we assume that the channel noise $\{w_k\}$ is an uncorrelated process with variance σ_w^2 that is uncorrelated with the source.

The theory of MMSE estimation is well established and widely accessible [see, e.g., Haykin (1996)]. Since we will find it convenient to refer to the geometrical aspects of MMSE estimation, especially later in our presentation of the MSE properties of the CM receiver, we introduce some of the basic concepts at this point. Minimizing MSE can be translated into the problem of finding the minimum Euclidean distance of a vector to a plane spanned by observations. Consider the Hilbert space defined by the joint probability distributions of all random variables (Caines 1988). By the orthogonality principle (Haykin 1996), the Wiener receiver's output, say $\bar{y}_n = \mathbf{r}^T(n)\mathbf{f}_m^{(\delta)}$, is obtained by projecting $s_{n-\delta}$ onto the subspace $\mathcal{Y}$ spanned by the observation contained in $\mathbf{r}(n)$, as shown in Fig. 2.4.

Figure 2.4 illustrates how, as an estimate of $s_{n-\delta}$, the Wiener output $\bar{y}_n$ is conditionally biased, that is,

$$\mathrm{E}\{\bar{y}_n | s_{n-\delta}\} \neq s_{n-\delta} \tag{2.28}$$

The conditionally unbiased estimate of $s_{n-\delta}$, denoted here by u_n, is given by scaling $\bar{y}_n$ such that its projection onto the direction of $s_{n-\delta}$ is $s_{n-\delta}$

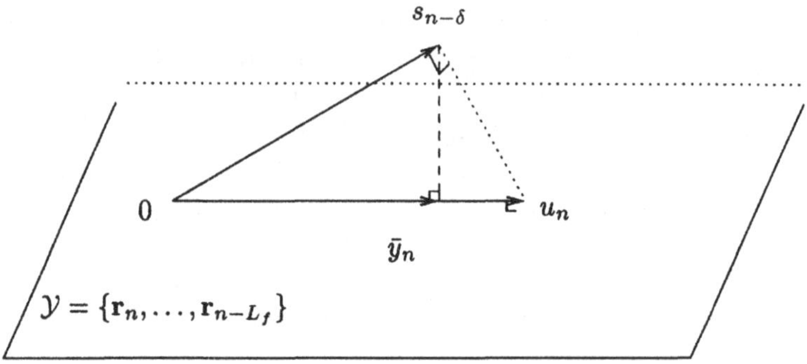

Figure 2.4 The geometrical interpretation of the Wiener estimator.

itself. It is important to note that SER performance is measured via u_n (rather than $\bar{y}_n$), and thus the comparison of the Wiener and CM receivers should also be made through the conditionally unbiased estimates. This idea will be revisited in Section 2.3.

For simplicity, we focus the remainder of the section on real-valued source, noise, channel, and equalizer quantities. Note, however, that the z-transforms of these real-valued quantities will be complex-valued.

The IIR Wiener Equalizer For the IIR equalizer, we assume for simplicity that $\delta = 0$ and drop the superscript notation on $\mathbf{f}_{\mathrm{m}}^{(0)}$. This simplification is justified by the fact that the MSE performance of the IIR equalizer is independent of delay choice δ. The (noncausal) IIR Wiener receiver $\mathbf{f}_{\mathrm{m}}$ can be derived most easily in the z-domain. By the orthogonality principle,

$$\mathrm{E}\left\{\mathbf{r}(z)\left(\mathbf{r}^T\left(\frac{1}{z^*}\right)\mathbf{f}_{\mathrm{m}}\left(\frac{1}{z^*}\right) - s\left(\frac{1}{z^*}\right)\right)^*\right\} = \mathbf{0} \tag{2.29}$$

Then, solving for $\mathbf{f}_{\mathrm{m}}(z)$,

$$\mathrm{E}\left\{\mathbf{r}(z)\mathbf{r}^H\left(\frac{1}{z^*}\right)\right\}\mathbf{f}_{\mathrm{m}}^*\left(\frac{1}{z^*}\right) = \mathbf{h}(z) \tag{2.30}$$

Finally, using $\mathbf{r}(z) = \mathbf{h}(z)s(z) + \mathbf{w}(z)$ and the matrix inversion lemma[5] (Kailath 1980),

[5] The matrix inversion lemma is commonly written as $\mathbf{A}^{-1} = (\mathbf{B}^{-1} + \mathbf{C}\mathbf{D}^{-1}\mathbf{C}^H)^{-1} = \mathbf{B} - \mathbf{B}\mathbf{C}(\mathbf{D} + \mathbf{C}^H\mathbf{B}\mathbf{C})^{-1}\mathbf{C}^H\mathbf{B}$, where $\mathbf{A}$ and $\mathbf{B}$ are positive definite $M \times M$ matrices, $\mathbf{D}$ is a positive definite $N \times N$ matrix, and $\mathbf{C}$ is an $M \times N$ matrix. In deriving Eq. (2.31) we use the inversion lemma to find $\mathbf{A}^{-1}$, where $\mathbf{A} = \mathrm{E}\{\mathbf{r}(z)\mathbf{r}^H(1/z^*)\}$, by choosing $\mathbf{B} = 1/\sigma_w^2$, $\mathbf{C} = \mathbf{h}(z)$, and $\mathbf{D} = 1/\sigma_s^2 = 1$.

$$\mathbf{f}_{\mathrm{m}}(z) = \frac{1}{\mathbf{h}^T(z)\mathbf{h}^*(1/z^*) + \sigma_w^2}\mathbf{h}^*\left(\frac{1}{z^*}\right) \tag{2.31}$$

where $\sigma_w^2 = \mathrm{E}\{|w_k|^2\}$ is the noise power. By setting $z = e^{j\omega}$, we obtain the frequency response of the Wiener equalizer:

$$\mathbf{f}_{\mathrm{m}}(e^{j\omega}) = \frac{1}{\|\mathbf{h}(e^{j\omega})\|^2 + \sigma_w^2}\mathbf{h}^*(e^{j\omega}) \tag{2.32}$$

The denominator term $\|\mathbf{h}(e^{j\omega})\|^2 = \sum_{p=1}^{P} |h^{(p)}(e^{j\omega})|^2$ is sometimes called the *folded-channel spectrum* (Lee and Messerschmitt 1994, 228). Recall that $h^{(p)}(e^{j\omega})$ is the frequency response of the pth subchannel.

System Response The combined channel-equalizer response $q_{\mathrm{m}}(z)$ resulting from IIR Wiener equalization is

$$q_{\mathrm{m}}(z) = \mathbf{f}_{\mathrm{m}}^T(z)\mathbf{h}(z) = \frac{\mathbf{h}^H(1/z^*)\mathbf{h}(z)}{\mathbf{h}^H(1/z^*)\mathbf{h}(z) + \sigma_w^2} \tag{2.33}$$

or in the frequency domain

$$q_{\mathrm{m}}(e^{j\omega}) = \frac{\|\mathbf{h}(e^{j\omega})\|^2}{\|\mathbf{h}(e^{j\omega})\|^2 + \sigma_w^2} \tag{2.34}$$

As $\sigma_w^2 \to 0$, $q_{\mathrm{m}}(z) \to 1$ as long as the subchannels have no common roots on the unit circle (i.e., $\forall\omega$, $\exists p, \ell$ s.t. $h^{(p)}(e^{j\omega}) \neq h^{(\ell)}(e^{j\omega})$). In this case, the folded spectrum has no nulls, and perfect symbol recovery is achieved.

MMSE of IIR Wiener Receiver Using Eq. (2.11), the estimation error of the IIR Wiener receiver is given by

$$e(z) = s(z) - y(z) = (1 - \mathbf{f}_{\mathrm{m}}^T(z)\mathbf{h}(z))s(z) - \mathbf{f}_{\mathrm{m}}^T(z)\mathbf{w}(z) \tag{2.35}$$

Then, with the assumption that $\sigma_s^2 = 1$, the power spectrum of the error sequence of the Wiener filter has the form

$$S_e(\omega) = |1 - \mathbf{f}_{\mathrm{m}}^T(e^{j\omega})\mathbf{h}(e^{j\omega})|^2 + \sigma_w^2\|\mathbf{f}_{\mathrm{m}}(e^{j\omega})\|^2 \tag{2.36}$$

$$= \frac{\sigma_w^2}{\|\mathbf{h}(e^{j\omega})\|^2 + \sigma_w^2} \tag{2.37}$$

The MSE $\mathscr{E}_{\mathrm{m}}$ of the Wiener filter is then given by

$$\mathscr{E}_{\mathrm{m}} = \frac{1}{2\pi}\int_{-\pi}^{\pi} \frac{\sigma_w^2}{\|\mathbf{h}(e^{j\omega})\|^2 + \sigma_w^2}\, d\omega \tag{2.38}$$

The FIR Wiener Equalizer When the vector equalizer polynomial $\mathbf{f}_{\mathrm{m}}^{(\delta)}(z)$ is of finite order, it is convenient to derive the Wiener equalizer in the time domain. From Fig. 2.4, the Wiener receiver satisfies the orthogonality principle: $(s_{n-\delta} - y_n) \perp \mathbf{r}(n)$, or more specifically,

$$\mathrm{E}\{\mathbf{r}(n)(s_{n-\delta} - y_n)\} = \mathbf{0} \tag{2.39}$$

Equation (2.39) leads, via Eq. (2.13), to the Wiener-Hopf equation whose solution is the Wiener receiver (Haykin 1996):

$$\mathbf{R}_{r,r}\mathbf{f}_{\mathrm{m}}^{(\delta)} = \mathbf{d}_{r,s}^{(\delta)} \Rightarrow \mathbf{f}_{\mathrm{m}}^{(\delta)} = \mathbf{R}_{r,r}^{\dagger}\mathbf{d}_{r,s}^{(\delta)} \tag{2.40}$$

Here $\mathbf{R}_{r,r} \triangleq \mathrm{E}\{\mathbf{r}(n)\mathbf{r}^T(n)\}$, $\mathbf{d}_{r,s}^{(\delta)} \triangleq \mathrm{E}\{\mathbf{r}(n)s_{n-\delta}\}$, and $(\cdot)^{\dagger}$ denotes the Moore-Penrose pseudoinverse (Strang 1988). In the case of mutually uncorrelated input and noise processes and $\sigma_s^2 = 1$, $\mathbf{R}_{r,r} = \mathbf{H}\mathbf{H}^T + \sigma_w^2\mathbf{I}$ and $\mathbf{d}_{r,s}^{(\delta)} = \mathbf{H}\mathbf{e}_{\delta}$. This yields the following expression for the FIR Wiener equalizer:

$$\mathbf{f}_{\mathrm{m}}^{(\delta)} = (\mathbf{H}\mathbf{H}^T + \sigma_w^2\mathbf{I})^{\dagger}\mathbf{H}\mathbf{e}_{\delta} \tag{2.41}$$

With the help of the singular-value decomposition (SVD) (Strang 1988), we can obtain an interpretation of the FIR Wiener receiver that is analogous to Eq. (2.33). Let $\mathbf{H}$ have the following SVD:

$$\mathbf{H} = \mathbf{U}\boldsymbol{\Sigma}\mathbf{V}^T \qquad \text{with} \qquad \boldsymbol{\Sigma} = \mathrm{diag}\{\varsigma_0, \ldots, \varsigma_{L_f+L_h}\} \tag{2.42}$$

where $\boldsymbol{\Sigma}$ has the same dimensions as $\mathbf{H}$ (which may not be square) and has the singular values $\{\varsigma_i\}$ on its first diagonal. We then have

$$\mathbf{f}_{\mathrm{m}}^{(\delta)} = \mathbf{U}(\boldsymbol{\Sigma}\boldsymbol{\Sigma}^T + \sigma_w^2\mathbf{I})^{\dagger}\boldsymbol{\Sigma}\mathbf{V}^T\mathbf{e}_{\delta} \tag{2.43}$$

The preceding formula resembles the frequency-domain solution given in Eqs. (2.31)–(2.32). In fact, Eq. (2.43) can be rewritten as

$$\mathbf{f}_{\mathrm{m}}^{(\delta)} = \mathbf{U}\,\mathrm{diag}\left\{\frac{\varsigma_0}{\varsigma_0^2 + \sigma_w^2}, \ldots, \frac{\varsigma_{L_f+L_h}}{\varsigma_{L_f+L_h}^2 + \sigma_w^2}\right\}\mathbf{V}^T\mathbf{e}_{\delta} \tag{2.44}$$

Note that the terms involving the singular values $\{\varsigma_i\}$ in Eq. (3.44) have a form reminiscent of the frequency response in Eq. (2.32).

System Response The Wiener system response is given by

$$\mathbf{q}_{\mathrm{m}}^{(\delta)} = \mathbf{H}^T\mathbf{f}_{\mathrm{m}}^{(\delta)} = \mathbf{H}^T(\mathbf{H}\mathbf{H}^T + \sigma_w^2\mathbf{I})^{\dagger}\mathbf{H}\mathbf{e}_\delta \tag{2.45}$$

To obtain a form similar to that of the IIR case, we use the SVD expressions, Eqs. (2.42) and (2.43), to obtain

$$\mathbf{q}_{\mathrm{m}}^{(\delta)} = \mathbf{V}\mathbf{\Sigma}^T(\mathbf{\Sigma}\mathbf{\Sigma}^T + \sigma_w^2\mathbf{I})^{\dagger}\mathbf{\Sigma}\mathbf{V}^T\mathbf{e}_\delta \tag{2.46}$$

where the singular values of the channel matrix play the role of magnitude spectrum of the channel in the IIR case. When $\mathbf{H}$ has column dimension less than or equal to its row dimension[6] column rank $\mathbf{H}$ we have

$$\mathbf{q}_{\mathrm{m}}^{(\delta)} = \mathbf{V}\,\mathrm{diag}\left\{\frac{\varsigma_0^2}{\varsigma_0^2 + \sigma_w^2}, \ldots, \frac{\varsigma_{L_f+L_h}^2}{\varsigma_{L_f+L_h}^2 + \sigma_w^2}\right\}\mathbf{V}^T\mathbf{e}_\delta \tag{2.47}$$

Note the similarity between Eq. (2.33) and Eq. (2.47). Again, as $\sigma_w^2 \to 0$, $\mathbf{q}_{\mathrm{m}}^{(\delta)} \to \mathbf{e}_\delta$, and perfect symbol recovery is achieved. In this case, the Wiener and ZF equalizers are identical ($\mathbf{f}_{\mathrm{m}}^{(\delta)} = \mathbf{f}_{\mathrm{z}}^{(\delta)}$). We note in advance that, for full column rank $\mathbf{H}$ (i.e., $\varsigma_i > 0$) and $\sigma_w^2 = 0$, CM receivers also achieve perfect symbol recovery up to a fixed phase ambiguity. On the other hand, by increasing $\sigma_w^2/\sigma_s^2 \to \infty$, $\mathbf{q}_{\mathrm{m}}^{(\delta)}$ approaches the origin. Interestingly, this property is *not* shared by the CM receivers, as we shall describe in Section 2.3.

MMSE of FIR Wiener Receiver For a given system delay δ and equalizer length $P(L_f + 1)$, the MMSE $\mathscr{E}_{\mathrm{m}}^{(\delta)}$ is defined as $J_{\mathrm{m}}^{(\delta)}(\mathbf{f}_{\mathrm{m}}^{(\delta)})$. A simplified expression can be obtained by substituting the Wiener expression, Eq. (2.41), into

$$\mathscr{E}_{\mathrm{m}}^{(\delta)} = \|\mathbf{H}^T\mathbf{f}_{\mathrm{m}}^{(\delta)} - \mathbf{e}_\delta\|_2^2 + \sigma_w^2\|\mathbf{f}_{\mathrm{m}}^{(\delta)}\|_2^2 \tag{2.48}$$

[6] If $\mathbf{H}$ has more columns than rows, that is, the equalizer is "undermodeled" with respect to the channel, then

$$\mathbf{q}_{\mathrm{m}}^{(\delta)} = \mathbf{V}\,\mathrm{diag}\left\{\frac{\varsigma_0^2}{\varsigma_0^2 + \sigma_w^2}, \ldots, \frac{\varsigma_{L_f+L_h}^2}{\varsigma_{L_f+L_h}^2 + \sigma_w^2}, 0, \ldots, 0\right\}\mathbf{V}^T\mathbf{e}_\delta$$

or can be obtained by first applying the orthogonality principle, Eq. (2.39), to the MSE definition, Eq. (2.26) and then substituting the expression (2.41) as follows:

$$\begin{aligned}\mathscr{E}_{\mathrm{m}}^{(\delta)} &= \mathrm{E}\{|\mathbf{r}^T(n)\mathbf{f}_{\mathrm{m}}^{(\delta)} - s_{n-\delta}|^2\} \\ &= \mathrm{E}\{-s_{n-\delta}(\mathbf{r}^T(n)\mathbf{f}_{\mathrm{m}}^{(\delta)} - s_{n-\delta})\} \\ &= 1 - \mathbf{e}_\delta^T\mathbf{H}^T(\mathbf{H}\mathbf{H}^T + \sigma_w^2\mathbf{I})^\dagger\mathbf{H}\mathbf{e}_\delta \end{aligned} \tag{2.49}$$

Effects of various parameters on the MMSE can be analyzed using the SVD. Substituting Eq. (2.42) into Eq. (2.49), we get

$$\mathscr{E}_{\mathrm{m}}^{(\delta)} = 1 - \sum_{i=0}^{L_f+L_h} \frac{\varsigma_i^2}{\varsigma_i^2 + \sigma_w^2}|v_{\delta,i}|^2 = \sum_{i=0}^{L_f+L_h} \frac{\sigma_w^2}{\varsigma_i^2 + \sigma_w^2}|v_{\delta,i}|^2 \tag{2.50}$$

where $v_{\delta,i}$ is the (δ, i)th entry of $\mathbf{V}$, that is, the δth entry of the ith right singular vector of $\mathbf{H}$.

From Eq. (2.50) we see that $\mathscr{E}_{\mathrm{m}}^{(\delta)}$ depends on four factors: (1) the noise power, (2) the subequalizer order L_f, (3) the singular values and singular vectors of the channel matrix $\mathbf{H}$, and (4) the system delay δ.

- *Effects of Noise.* As $\sigma_w^2 \to 0$, $\mathscr{E}_{\mathrm{m}}^{(\delta)}$ decreases, but not necessarily to zero. The only case in which $\mathscr{E}_{\mathrm{m}}^{(\delta)}$ approaches zero (for all δ) is when $\varsigma_i > 0$, that is, $\mathbf{H}$ has full column rank. For nonzero σ_w^2, Eq. (2.48) indicates that the MMSE equalizer achieves a compromise between noise gain (i.e., $\|\mathbf{f}_{\mathrm{m}}^{(\delta)}\|_2$) and ISI (i.e., $\|\mathbf{q}_{\mathrm{m}}^{(\delta)} - \mathbf{e}_\delta\|_2$).
- *Subequalizer Order L_f.* $\mathscr{E}_{\mathrm{m}}^{(\delta)}$ is a nonincreasing function of L_f. Using the Toeplitz distribution theorem (Gray 1972), it can be shown that the FIR MMSE, Eq. (2.50), approaches the IIR MMSE, Eq. (2.38), as $L_f \to \infty$, which is intuitively satisfying. In practice, the selection of L_f leads to a trade-off between desired performance and implementation complexity.
- *Effects of Channel.* For FIR equalizers, Eq. (2.50) suggests a relatively complex relationship between MMSE and the singular values and right singular vectors of the channel matrix. In the case of IIR equalizers, there exists a much simpler relationship between channel properties and MMSE performance. Specifically, Eq. (2.38) indicates that common subchannel roots near the unit circle cause an increase in MMSE. Though the relationship between MMSE and subchannel roots is less obvious in the case of a FIR equalizer, it

has been shown that increasing the proximity of subchannel roots decreases the product of the singular values [see Fijalkow (1996) and Casas et al. (1997)]. Furthermore, the effects of near-common subchannel roots are more severe when noise power is large. Unfortunately, however, a direct link between subchannel root locations and FIR MMSE has yet to be found.

- *System Delay* δ. For FIR Wiener receivers, selection of δ can affect MMSE significantly. This can be seen in Eq. (2.49), where the δth diagonal element of the matrix quantity $\mathbf{H}^T(\mathbf{H}\mathbf{H}^T + \sigma_w^2\mathbf{I})^{\dagger}\mathbf{H}$ is extracted by the $\mathbf{e}_\delta$ pair. Figure 2.22 in Section 2.4 shows an example of $\mathcal{E}_{\mathrm{m}}^{(\delta)}$ for various equalizer lengths. Typically, a low-MSE "trough" exists for system delays in the vicinity of the channel's center of gravity, and system delays outside of this trough exhibit markedly higher MMSE. This can be contrasted to the performance of the IIR Wiener receiver that is invariant to delay choice. We note in advance that system delay is not a direct design parameter with CMA-adapted FIR equalizers (as discussed in Section 2.4).

2.2.5 The LMS Algorithm

The LMS algorithm (Widrow and Stearns 1985) is one of the most widely used stochastic-gradient descent (SGD) algorithms for adaptively minimizing MSE. In terms of the instantaneous squared error

$$\hat{J}_{\mathrm{m}}^{(\delta)}(n) \triangleq \tfrac{1}{2}|y_n - s_{n-\delta}|^2 \tag{2.51}$$

the real-valued LMS parameter-vector update equation is

$$\mathbf{f}(n+1) = \mathbf{f}(n) - \mu\nabla_{\mathbf{f}}\hat{J}_{\mathrm{m}}^{(\delta)}(n) \tag{2.52}$$

$$= \mathbf{f}(n) - \mu\mathbf{r}(n)(y_n - s_{n-\delta}) \tag{2.53}$$

where $\nabla_{\mathbf{f}}$ denotes the gradient with respect to the equalizer coefficient vector and μ is a (small) positive step size. In practice, a training sequence sent by the transmitter and known *a priori* by the receiver is used to supply the $s_{n-\delta}$ term in Eq. (2.53).

Standard analysis of LMS considers the transient and steady-state properties of the algorithm separately. Detailed expositions on LMS can be found in, for example, Haykin (1996), Widrow and Stearns (1985), Gitlin et al. (1992), and Macchi (1995). We shall review the basic behavior of the LMS algorithm (e.g., excess MSE and convergence rate) in the equalization context for later comparison with the CM-minimizing algorithm CMA. Many similarities can be found between LMS and CMA because both attempt to minimize their respective costs ($J_{\mathrm{m}}^{(\delta)}$ and J_{cm}) using an SGD technique.

Transient Behavior The principal item of interest in the transient behavior of LMS is convergence rate. Because the MSE cost $J_{\mathrm{m}}^{(\delta)}$ is quadratic, the Hessian is constant throughout the parameter space and thus convergence rate analysis is straightforward. (The Hessian matrix, defined as $\nabla_{\mathbf{f}}\nabla_{\mathbf{f}^T}J_{\mathrm{m}}^{(\delta)}$, determines the curvature of the cost surface.) Below we derive bounds on the convergence rate of a FIR equalizer.

The instantaneous error can be partitioned into three components: the noise and ISI contributions to Wiener equalization and the error resulting from any deviations from Wiener equalization:[7]

$$e_n = y_n - s_{n-\delta} \tag{2.54}$$

$$= \mathbf{r}^T(n)\mathbf{f}(n) - \mathbf{x}^T(n)\mathbf{f}_{\mathrm{z}}^{(\delta)} \tag{2.55}$$

$$= \underbrace{\mathbf{r}^T(n)(\mathbf{f}(n) - \mathbf{f}_{\mathrm{m}}^{(\delta)})}_{\text{from parameter errors}} + \underbrace{\mathbf{x}^T(n)(\mathbf{f}_{\mathrm{m}}^{(\delta)} - \mathbf{f}_{\mathrm{z}}^{(\delta)})}_{\text{from ISI and bias}} + \underbrace{\mathbf{w}^T(n)\mathbf{f}_{\mathrm{m}}^{(\delta)}}_{\text{from noise}} \tag{2.56}$$

In deriving Eq. (2.56), we used the definition $\mathbf{x}(n) = \mathbf{H}\mathbf{s}(n)$ and the ZF equalizer property $\mathbf{H}^T\mathbf{f}_{\mathrm{z}}^{(\delta)} = \mathbf{e}_\delta$ that appeared earlier in this section.

We define the equalizer *parameter error* vector as $\tilde{\mathbf{f}}(n) \triangleq \mathbf{f}(n) - \mathbf{f}_{\mathrm{m}}^{(\delta)}$. Substituting Eq. (2.56) into the LMS update equation (2.53) and subtracting $\mathbf{f}_{\mathrm{m}}^{(\delta)}$ from both sides, the parameter error can be seen to evolve as

$$\tilde{\mathbf{f}}(n+1) = (\mathbf{I} - \mu\mathbf{r}(n)\mathbf{r}^T(n))\tilde{\mathbf{f}}(n) - \mu\mathbf{r}(n)\mathbf{x}^T(n)(\mathbf{f}_{\mathrm{m}}^{(\delta)} - \mathbf{f}_{\mathrm{z}}^{(\delta)}) - \mu\mathbf{r}(n)\mathbf{w}^T(n)\mathbf{f}_{\mathrm{m}}^{(\delta)} \tag{2.57}$$

With a reasonably small step-size, the parameters change very slowly with respect to the signal and noise vectors in Eq. (2.56). To exploit this dual time-scale nature of the LMS update, we define the average parameter error vector $\mathbf{g}(n) \triangleq \mathrm{E}\{\tilde{\mathbf{f}}(n)\}$. Then, assuming that the signal and noise are mutually uncorrelated and denoting the autocorrelation matrix of the (real-valued) noiseless received signal by $\mathbf{R}_{x,x} \triangleq \mathrm{E}\{\mathbf{x}(n)\mathbf{x}^T(n)\}$,

[7] In Eq. (2.56), $\mathbf{f}(n)$ is allowed to be of arbitrary length. While a Wiener equalizer always exists to match the length of $\mathbf{f}(n)$, the ZF equalizer $\mathbf{f}_{\mathrm{z}}^{(\delta)}$ must, in general, satisfy the length condition $L_f \geq L_h - 1$. Thus, in Eq. (2.56), $\mathbf{f}_{\mathrm{z}}^{(\delta)}$ must be zero-padded when $L_f > L_h - 1$, while the $\mathbf{f}_{\mathrm{m}}^{(\delta)}$ in the "from ISI and bias" term must be zero padded when $L_f < L_h - 1$.

Eq. (2.57) implies that the average parameter error evolves as

$$\mathbf{g}(n+1) = (\mathbf{I} - \mu\mathbf{R}_{r,r})\mathbf{g}(n) - \mu\mathbf{R}_{x,x}(\mathbf{f}_{\mathrm{m}}^{(\delta)} - \mathbf{f}_{\mathrm{z}}^{(\delta)}) - \mu\sigma_w^2\mathbf{f}_{\mathrm{m}}^{(\delta)} \tag{2.58}$$

We have utilized the assumption that $\mathbf{f}(n)$ is changing slowly with respect to the data $\mathbf{r}(n)$, so that the average of their products can be well approximated by the product of their averages.

Our expression for average parameter error can be simplified with the observation that

$$\begin{aligned}\mathbf{R}_{x,x}(\mathbf{f}_{\mathrm{m}}^{(\delta)} - \mathbf{f}_{\mathrm{z}}^{(\delta)}) + \sigma_w^2\mathbf{f}_{\mathrm{m}}^{(\delta)} &= \mathbf{R}_{r,r}\mathbf{f}_{\mathrm{m}}^{(\delta)} - \mathbf{R}_{x,x}\mathbf{f}_{\mathrm{z}}^{(\delta)} = \mathbf{d}_{r,s}^{(\delta)} - \mathbf{H}\mathbf{H}^T\mathbf{f}_{\mathrm{z}}^{(\delta)} \\ &= \mathbf{d}_{r,s}^{(\delta)} - \mathbf{H}\mathbf{e}_\delta = \mathbf{0}\end{aligned}$$

giving the LMS average parameter update equation

$$\mathbf{g}(n+1) = (\mathbf{I} - \mu\mathbf{R}_{r,r})\mathbf{g}(n) \tag{2.59}$$

Notice that Eq. (2.59) specifies a linear homogeneous difference equation with state transition matrix $(\mathbf{I} - \mu\mathbf{R}_{r,r})$. For stability of the average-error-system Eq. (2.59), the eigenvalues of $(\mathbf{I} - \mu\mathbf{R}_{r,r})$ must lie between -1 and 1. Letting $\lambda_{\max}$ be the maximum eigenvalue of $\mathbf{R}_{r,r}$, this implies that the stability of the average system is guaranteed when

$$0 < \mu < \frac{2}{\lambda_{\max}} \tag{2.60}$$

A classic implication of Eq. (2.59) is that, for μ obeying Eq. (2.60), the average LMS parameter error decays to zero as the equalizer adapts. In other words, LMS converges in mean to the Wiener solution $\mathbf{f}_{\mathrm{m}}^{(\delta)}$.

Mean-*square* convergence requires, in addition to Eq. (2.60), that (Haykin 1996)

$$\sum_i \frac{\mu\lambda_i}{2 - \mu\lambda_i} < 1$$

which, for small step sizes, can be approximated by

$$\mu < \frac{2}{\sum_i \lambda_i} = \frac{2}{P(L_f + 1)\sigma_r^2} \tag{2.61}$$

Equation (2.61) indicates a step-size requirement proportional to equalizer length and received-signal power.

Asymptotic Performance When any of the perfect symbol-recovery conditions is violated, no fixed equalizer can zero the error $e_n = y_n - s_{n-\delta}$ used by the LMS algorithm. Thus, with a nonvanishing step size μ, an LMS-derived parameter vector $\mathbf{f}(n)$ will never settle to the MMSE solution, but instead "jitter" around it. Therefore, the actual steady-state MSE achieved by LMS is greater than the MMSE. We call the difference between MSE and MMSE the *excess mean square error* (EMSE).

EMSE can be approximated by a function of step-size and equalizer length. Using Eq. (2.56) and the assumptions that $\mathbf{x}(n), \mathbf{w}(n)$, and $\tilde{\mathbf{f}}(n)$ are uncorrelated, we can obtain an expression for the steady-state MSE in terms of EMSE and MMSE (Haykin 1996):

$$\underbrace{\mathrm{E}\{e_n^2\}}_{\text{MSE}} = \underbrace{\mathrm{tr}(\mathbf{R}_{r,r}\mathrm{E}\{\tilde{\mathbf{f}}(n)\tilde{\mathbf{f}}^T(n)\})}_{\text{EMSE}, \mathscr{E}_\chi} + \underbrace{\|\mathbf{f}_{\mathrm{m}}^{(\delta)} - \mathbf{f}_{\mathrm{z}}^{(\delta)}\|_{\mathbf{R}_{x,x}}^2 + \sigma_w^2 \|\mathbf{f}_{\mathrm{m}}^{(\delta)}\|_2^2}_{\text{MMSE}, \mathscr{E}_{\mathrm{m}}} \tag{2.62}$$

The EMSE term $\mathscr{E}_\chi$ is, in general, difficult to analyze. However, for small μ, a close approximation to $\mathscr{E}_\chi$ can be found using various independence assumptions on $\mathbf{r}(n)$, $\mathbf{s}(n)$, and $\tilde{\mathbf{f}}(n)$. In that case,

$$\mathscr{E}_\chi \approx \frac{\mu}{2} P(L_f + 1)\sigma_r^2 \mathscr{E}_{\mathrm{m}} \tag{2.63}$$

where $\sigma_r^2 \triangleq \mathrm{E}\{|r_k|^2\}$ is the received signal power (Haykin 1996). Note that EMSE is proportional to the product of step-size, equalizer length, received signal power, and MMSE, Recall that $P(L_f + 1)$ represents the total number of adapted equalizer coefficients.

LMS Design Implications The form of the EMSE expression, Eq. (2.63), has a number of design implications for LMS-based adaptive equalization.

For one, there is the step-size trade-off: a large step size gives fast convergence to an equalizer with large EMSE, while a small step size gives slow convergence to an equalizer with small EMSE. This suggests making μ as small as possible within allowed limits on convergence time. When the channel impulse response is time varying, however, the situation becomes more complicated: step sizes that are too small may not be able to adequately track the channel variations, resulting in what is known as *tracking lag* (Widrow et al. 1976). Tracking lag prevents the equalizer from remaining in close vicinity to the time-varying $\mathbf{f}_{\mathrm{m}}^{(\delta)}$, which ultimately increases MSE. Increasing the step size may reduce tracking lag, but may increase $\mathscr{E}_\chi$ via Eq. (2.63). Thus, even in time-varying situations, step sizes that are too large ultimately increase steady-state MSE via Eq. (2.62). Hence, the best step size is a compromise between these situations.

There is a similar trade-off with equalizer length $P(L_f + 1)$. Equalizers that are too short result in a high MMSE, which prohibits a low steady-state MSE via Eq. (2.62). On the other hand, very long equalizers may result in a large EMSE via Eq. (2.63), also preventing good steady-state MSE performance. Thus, the optimal equalizer length lies somewhere in between.

The analysis of Section 2.3 will demonstrate that CMA shares many of the same behavior features of LMS. Hence, many of the CMA design guidelines in Section 2.4 can be related back to the LMS design guidelines discussed in this section.

2.3 THE CM CRITERION AND CMA

This section focuses on the properties of the CM criterion and the behavior of the CMA. What will be a recurring theme can be summarized by *Godard's conjecture*, an observation made by Godard in his seminal paper on blind adaptive (BS) equalization. In Godard (1980), he observed that the MSE performance of CMA is close to that of the MMSE-optimal (Wiener) equalizer: "It should also be noted that the equalizer coefficients minimizing the dispersion functions closely approximate those which minimize the mean-squared error." Treichler and Agee (1983) made a similar claim in their independent development of the CM criterion. Throughout this section, we shall provide evidence supporting Godard's conjecture.

This section is organized as follows. First we introduce the CM criterion and present what are known as the perfect blind equalization (PBE) conditions. Simple illustrated examples are provided to understand the effects of violating these conditions. Next, properties of the local and global minima of the CM cost function (referred to as *CM receivers*) are discussed and compared to that of the well-known Wiener and ZF receivers discussed in Section 2.2. This is followed by summaries of important analytical work concerning the robustness of CM receivers to the presence of noise and channel undermodeling. Both perturbation-based approximations and bounds from a geometrical approach are used to predict CM performance in a practical setting. The section concludes by summarizing the transient and asymptotic behavior of CMA, an SGD algorithm popularly employed to adapt blind equalizers, which minimizes the CM criterion. Various comparisons are drawn between CMA and the well-known LMS algorithm (discussed in Section 2.2).

2.3.1 Properties of the CM Criterion

Functional Form Blind equalization can be considered as the estimation of (unknown) source symbols $\{s_n\}$ from the receiver input sequence $\{\mathbf{r}_n\}$, or equivalently $\{\mathbf{r}(n)\}$. In our formulation of the problem, the linear filter $\mathbf{f}$ generates the symbol estimate $y_n = \mathbf{f}^T\mathbf{r}(n)$, and we desire that $y_n \approx s_{n-\delta}$ for some fixed-integer delay δ.

In motivating the CM criterion, we first consider the family of Bussgang techniques used in blind equalization (Bellini 1994). Some intuition behind these Bussgang techniques comes from a consideration of trained and decision-directed LMS-based equalizer adaptation, described by the update equation

$$\mathbf{f}(n+1) = \mathbf{f}(n) - \mu\mathbf{r}^*(n)e_n \tag{2.64}$$

In the case of a trained update, $e_n = y_n - s_{n-\delta}$, and in the case of a decision-directed (DD) update, $e_n = y_n - \hat{s}_{n-\delta}$, where $\hat{s}_{n-\delta}$ is the symbol estimate generated by the decision device. Since high symbol-error rates prevent the success of DD equalization in cold start-up applications, we are motivated to find a more reliable error signal e_n. Consider instead applying a memoryless nonlinearity $g(\cdot)$ to the equalizer output and forming the error

$$e_n = y_n - g(y_n) \tag{2.65}$$

An update based on Eq. (2.65) would satisfy our criterion for *blind* equalization since e_n can be constructed solely from the received signal. Such algorithms are referred to as "Bussgang" (Bussgang 1952) and achieve equilibrium when the equalizer output satisfies $\mathrm{E}\{\mathbf{r}(n)y_n\} = \mathrm{E}\{\mathbf{r}(n)g(y_n)\}$ [via Eqs. (2.64) and (2.65)].

In the reception of complex source alphabets such as quadrature amplitude modulation (QAM), the receiver must not only remove ISI, but also synchronize its demodulator to the carrier frequency and phase employed by the transmitter. In a blind scenario (i.e., in the absence of a pilot signal), carrier synchronization *before* equalization (i.e., in a non-decision-directed mode) is especially difficult with large QAM constellations. Similarly, blind equalization before carrier synchronization is often a prohibitively difficult task: to the typical blind symbol estimation algorithm, carrier offset appears as a (potentially rapid) time variation in the channel response. Rapid time-variations are known to hinder the convergence of SGD algorithms such as Eq. (2.64), which rely on some degree of (cyclo)stationarity for proper operation. Thus, practical considerations motivate the decoupling of blind equalization from carrier recovery (Treichler et al. 1998).

The search for a carrier-phase independent algorithm led Godard to consider blind equalization techniques based only on the signal modulus $|y_n|$. He proposed the minimization of the following "dispersion" cost (Godard 1980):

$$\begin{aligned} J_p &\triangleq \frac{1}{2p}\mathrm{E}\{(|y_n|^p - \gamma)^2\} \\ &= \frac{1}{2p}\mathrm{E}\{(|\mathbf{f}^T\mathbf{r}(n)|^p - \gamma)^2\} \end{aligned} \tag{2.66}$$

where the real constant γ is chosen as a function of the source alphabet and of the integer p. Specifically, Godard showed that for sub-Gaussian sources (described in **(A4)** below), J_p is minimized by system responses generating zero ISI, and he showed that choosing $\gamma = \mathrm{E}\{|s_n|^{2p}\}/\mathrm{E}\{|s_n|^p\}$ ensures that local minima of J_p exist at the perfectly equalizing system responses (Godard 1980). In the case of $p = 2$, we refer to Eq. (2.66) as the CM cost or the *CM criterion*, and denote it by J_{cm}:

$$J_{\mathrm{cm}} \triangleq \frac{1}{4}\mathrm{E}\left\{\left(|y_n|^2 - \frac{\mathrm{E}\{|s_n|^4\}}{\sigma_s^2}\right)^2\right\} = \frac{1}{4}\mathrm{E}\left\{\left(|\mathbf{f}^T\mathbf{r}(n)|^2 - \frac{\mathrm{E}\{|s_n|^4\}}{\sigma_s^2}\right)^2\right\} \tag{2.67}$$

Equation (2.67) shows that the CM criterion penalizes dispersion of the squared output modulus $|y_n|^2$ away from the constant $(\mathrm{E}\{|s_n|^4\}/\sigma_s^2)$. As a phase-independent criterion for blind source recovery, minimizing dispersion seems intuitively satisfying for CM sources.[8] Remarkably, the CM criterion works almost as well with non-CM sources. We note that the CM idea was independently proposed for constant-modulus source sequences (i.e., $\exists \rho \in \mathbb{R}$ s.t. $\forall n, |s_n| = \rho$) by Treichler and Agee (1983).

[8] Sources derived from M-PSK constellations are examples of CM sources, since every symbol in the M-PSK alphabet has the same magnitude (or "modulus"). M-QAM sources for $M > 4$ do not have the CM property, since alphabet members differ in magnitude as well as phase. See Fig. 2.19 for an illustration.

It is important to note that Godard's blind equalization algorithm belongs to the Bussgang class. In fact, it can be shown to approximate the conditional-mean estimator of $|s_n|^2$ given $|y_n|^2$ (Bellini 1994), a valid estimator in the absence of carrier-phase information (or equivalently, when all rotations of the source constellation are assumed equally likely).

Having introduced the CM criterion as a phase-independent means of blind equalization (thus intended for complex-valued signals), it may seem strange that in the sequel we restrict our analysis to the case of real-valued quantities. We justify our position by claiming that nearly all of the intuition concerning the behavior of the CM criterion can be gained from the study of its real-valued incarnation with the benefit of a simplified presentation. We will attempt to note any situations in which complex-valued quantities lead to meaningful conceptual differences. With this in mind, we present a useful expansion of the CM cost in terms of the equalizer coefficients $\mathbf{f}$ for real-valued channels, real-valued i.i.d. noise, and real-valued i.i.d. sources. [See Johnson (1998) for derivation of this and more general CM cost expressions.]

$$\begin{aligned} J_{\text{cm}} &= \tfrac{1}{4}\sigma_s^4(\kappa_s - 3)\|\mathbf{H}^T\mathbf{f}\|_4^4 + \tfrac{3}{4}\sigma_s^4\|\mathbf{H}^T\mathbf{f}\|_2^4 + \tfrac{1}{4}\sigma_w^4(\kappa_w - 3)\|\mathbf{f}\|_4^4 \\ &\quad + \tfrac{3}{4}\sigma_w^4\|\mathbf{f}\|_2^4 + \tfrac{3}{2}\sigma_s^2\sigma_w^2\|\mathbf{H}^T\mathbf{f}\|_2^2\|\mathbf{f}\|_2^2 \\ &\quad - \tfrac{1}{2}\sigma_s^2\kappa_s(\sigma_s^2\|\mathbf{H}^T\mathbf{f}\|_2^2 + \sigma_w^2\|\mathbf{f}\|_2^2) + \tfrac{1}{4}\sigma_s^4\kappa_s^2 \end{aligned} \tag{2.68}$$

In the preceding equation, κ_s refers to the normalized kurtosis of the source process, defined below in Eq. (2.69). By analogy, κ_w denotes the kurtosis of the channel noise process $\{w_k\}$. The quantities σ_s^2 and σ_w^2 denote the source and noise variances, respectively. Note that, in the absence of noise, J_{cm} is a quartic function of the ℓ_4 and ℓ_2 norms of the channel-equalizer impulse response $\mathbf{H}^T\mathbf{f}$. The addition of noise brings a quartic dependence on the ℓ_4 and ℓ_2 norms of the equalizer impulse response $\mathbf{f}$.

As we shall see in the following sections, the local and global minimizers of J_{cm}, that is, the CM receivers $\mathbf{f}_c$, are of key importance. For example, they represent the asymptotic mean-convergence points of the constant modulus algorithm.

Perfect Blind Equalizability Conditions The set of conditions under which all minimizers of J_{cm} accomplish perfect symbol recovery are known as the *perfect blind equalization* (PBE) conditions (Fijalkow 1994; Foschini 1985; Li and Ding 1996). They are given below.

(A1) Full-column-rank channel matrix **H**;
(A2) No additive channel noise;
(A3) Symmetric i.i.d.[9] source, circularly symmetric ($\mathrm{E}\{s_n^2\} = 0$) in the complex-valued case;
(A4) Sub-Gaussian source: the normalized source kurtosis satisfies $\kappa_s < \kappa_g$.

The normalized source kurtosis κ_s is defined as

$$\kappa_s \triangleq \frac{\mathrm{E}\{|s_n|^4\}}{\sigma_s^4} \tag{2.69}$$

where the kurtosis of a real-valued Gaussian random process is $\kappa_g = 3$ and the kurtosis of a proper complex-valued Gaussian random process is $\kappa_g = 2$. Note that Conditions **(A1)** and **(A2)** pertain to the channel-equalizer pair's ability to achieve perfect equalization, as discussed in Section 2.2. Conditions **(A3)** and **(A4)** pertain solely to blind equalization based on the CM criterion.

Illustrative Examples of CM Cost Surface Deformations This section studies changes to the "shape" of the CM cost surface as a means of understanding the effect of various violations of the PBE conditions. Restricting our focus to the case of a real-valued two-tap $T/2$-spaced equalizer permits illustration of the cost surface as a function of equalizer parameters. We do not intend a rigorous analysis of CM robustness properties here—that will be the subject of later subsections.

Perfect Blind Equalizability We first consider a 4-tap channel and 2-tap equalizer, both $T/2$-spaced (i.e., $P = 2$), satisfying the column rank condition **(A1)**. Associated with this model are the following channel convolution matrix and equalizer coefficient vector:

$$\mathbf{H} = \begin{pmatrix} h_1 & h_3 \\ h_0 & h_2 \end{pmatrix}, \qquad \mathbf{f} = \begin{pmatrix} f_0 \\ f_1 \end{pmatrix}$$

A square invertible **H** ensures unique zero-forcing solutions to $\pm\mathbf{e}_\delta = \mathbf{H}^T\mathbf{f}$ for $\delta \in \{0, 1\}$. As discussed in Section 2.2, the existence of this in-

[9] Examination of the derivation in Johnson et al. (1998) reveals that the CM cost expression yielding the global convergence properties requires only fourth-order statistical independence of the source process.

Table 2.1 Summary of Channels Used for Two-Tap FSE Examples

Name	$T/2$-spaced Impulse Response	Classification
$\mathbf{h}_a$	$(-0.0901, 0.6853, 0.7170, -0.0901)^T$	Well-behaved
$\mathbf{h}_b$	$(1.0, -0.5, 0.2, 0.3)^T$	Well-behaved
$\mathbf{h}_c$	$(-0.0086, 0.0101, 0.9999, -0.0086)^T$	Nearly common subchannel roots
$\mathbf{h}_d$	$(1.0, -0.5, 0.2, 0.3, -0.2, -0.15)^T$	Undermodeled

verse requires that the even and odd subchannels, $h^{(1)}(z) = h_0 + h_2 z^{-1}$ and $h^{(2)}(z) = h_1 + h_3 z^{-1}$, respectively, do not share the same root location. Examples of 4-tap channels satisfying this non-common-root condition are given by $\mathbf{h}_a$ and $\mathbf{h}_b$[10] in Table 2.1.

For an i.i.d. source with kurtosis $\kappa_s = 1$ in the absence of channel noise [thus satisfying **(A2)**–**(A4)**], the CM cost is shown in Fig. 2.5 and 2.6 as a function of equalizer coefficients. In all upcoming contour plots, a "$*$" indicates an MMSE solution corresponding to an optimal system delay, while a "$\times$" indicates an MMSE solution corresponding to a suboptimal system delay. In Fig. 2.6, however, zero MSE can be achieved at all system delays, and thus all solutions are optimal. It can be seen that when the PBE conditions are satisfied, the MMSE solutions exactly coincide with the minima of J_{cm}. Each pair of CM minima symmetric with respect to the origin corresponds to the same system delay but to two different[11] choices of system polarity ($\pm$). Such sign ambiguity is inconsequential when, for example, the symbols have been differentially encoded. Thus, in Fig. 2.5, the four minima correspond to the four permutations of system delay ($\delta = 0, 1$) and polarity ($+, -$). Note the CM local maximum at the origin.

Sometimes it is more convenient to study J_{cm} as a function of system response $\mathbf{q}$. Note that when $\mathbf{H}$ is square and invertible, as in our current example, there exists a unique mapping between $\mathbf{f}$- and $\mathbf{q}$-spaces: $\mathbf{q} = \mathbf{H}^T\mathbf{f}$. One appeal of studying J_{cm} in $\mathbf{q}$-space follows from the normalization and alignment of the J_{cm} minima with the coordinate axes. For example, our current ZF system responses occur at $\mathbf{q} = (\pm 1, 0)$ and $(0, \pm 1)$ (see Fig. 2.7). Such symmetry will be exploited in this section.

[10] $\mathbf{h}_a$ and h_b are deemed "well-behaved" to indicate the absence of common or nearly common subchannel roots.

[11] In the complex-valued case, each pair of minima would be replaced by a continuum of minima spanning the full range $(0 - 2\pi)$ of allowable system phase.

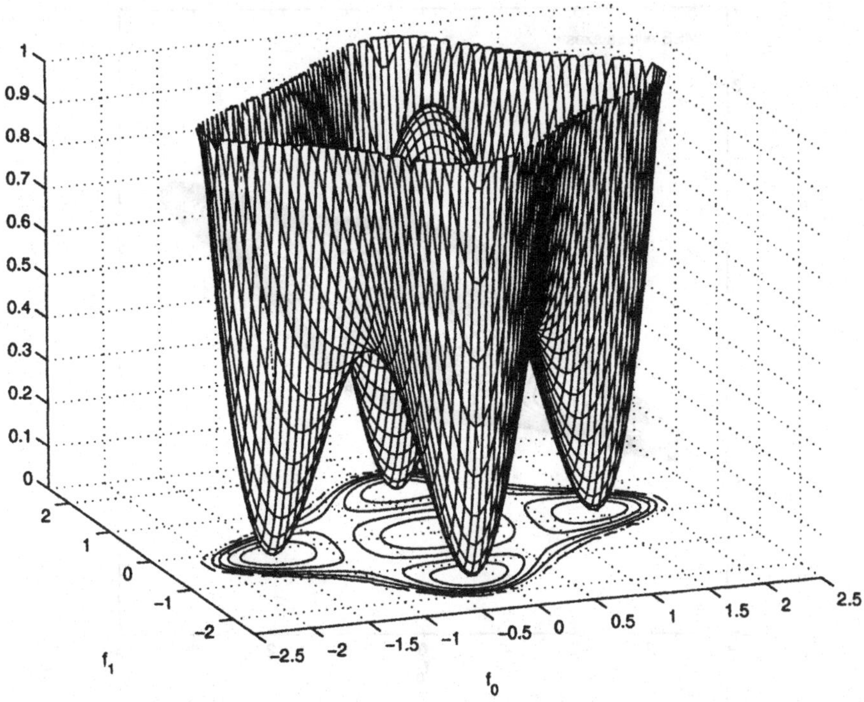

Figure 2.5 J_{cm} for BPSK, well-behaved channel $\mathbf{h}_a$, and no noise, in equalizer ($\mathbf{f}$) space.

Under satisfaction of **(A1)** and **(A2)**, viewing J_{cm} from system space has the additional advantage that the CM cost is channel independent [recall Eq. (2.68) with $\mathbf{H}^T\mathbf{f} = \mathbf{q}$].

Effect of Channel Noise Violation of **(A2)** occurs in any communication system where additive channel noise $\mathbf{w}(n)$ is present. We denote the signal-to-noise ratio (SNR) by $\text{SNR} = 10\log_{10}(\sigma_s^2/\sigma_w^2)$ using the source and noise variances introduced in Section 2.2. The presence of noise causes the shape of the CM cost surface to change in such a way that the CM minima no longer correspond to the Wiener minima and such that different CM minima may result in different levels of performance. In other words, we now have strictly local (versus global) minima. This can be recognized in the contour depths in Fig. 2.8.

Most importantly, however, Fig. 2.8 suggests that, even at relatively high noise powers, the CM and MSE minima roughly correspond. Here we see evidence of Godard's conjecture. It is interesting to note that the

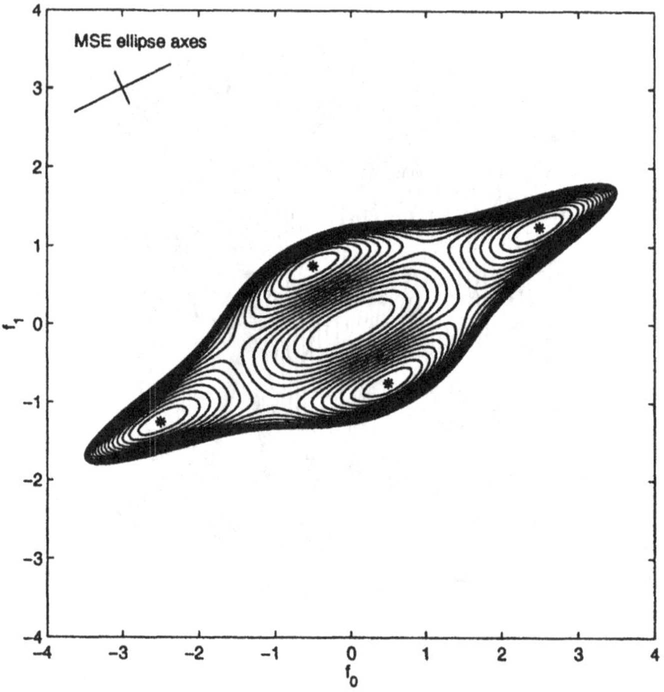

Figure 2.6 J_{cm} contours for BPSK, well-behaved channel $\mathbf{h}_b$, and no noise, in equalizer ($\mathbf{f}$) space, with global MMSE minima marked by $*$.

suboptimal CM minima are further from their neighboring Wiener solutions than the optimal CM minima are from theirs. The robustness properties of the CM criterion to additive channel noise are the focus of sections presented in the sequel.

Effect of Nearly Common Subchannel Roots In Section 2.2 we discussed the effects of (exactly) common subchannel roots and demonstrated that, in general, they prevent perfect source recovery [hence violating **(A1)**]. Though nearly common subchannel roots do not share this problem, they may excite an undesirable phenomenon known as *noise gain*, a familiar concept from ZF and MMSE equalization theory. We will see that a similar phenomenon exists for equalization methods based on minimizing the CM criterion.

One way to understand the mechanics of noise gain is to consider the transformation from the ZF system responses $\{\mathbf{q}_z^{(\delta)}\}$ to the respective ZF equalizers $\{\mathbf{f}_z^{(\delta)}\}$. Nearly common subchannel roots (i.e., $h_3/h_1 \approx h_2/h_0$) result in a transformation matrix $(\mathbf{H}^T)^\dagger$ with large eigenvalues. This has the effect of mapping certain ZF system responses to ZF equalizers with

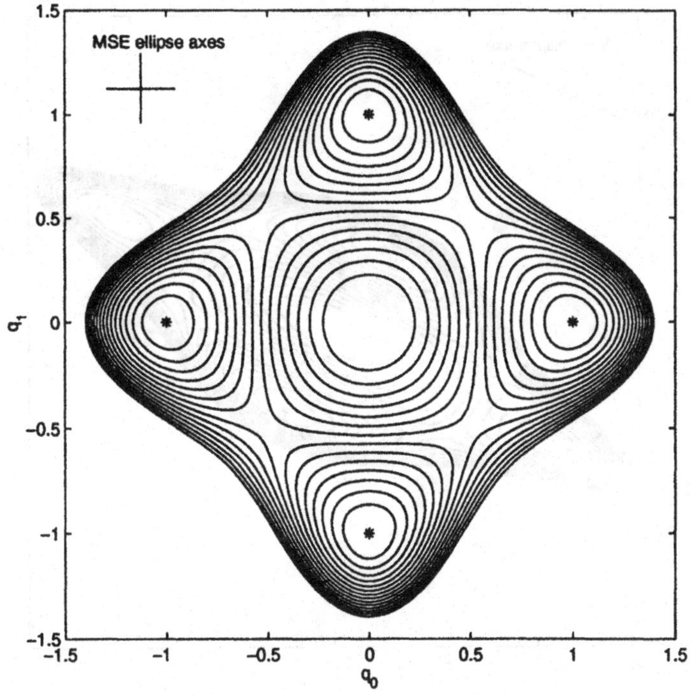

Figure 2.7 J_{cm} contours for a BPSK source in the absence of noise, in system (**q**) space.

large norm, as illustrated[12] by Fig. 2.9. These large-norm equalizers result in significant noise gain, as indicated by the MSE formula (2.48). In order to better compromise between ISI cancellation and noise gain, we expect the MMSE optimal equalizers to be smaller in norm than their ZF counterparts. This is confirmed by the behavior of the Wiener expressions (2.32) and (2.44) as σ_w^2 is increased. Equation (2.44) also indicates that if $\mathbf{H}$ has small singular values (as would result from nearly common subchannel roots), $\mathbf{f}_{\mathrm{m}}^{(\delta)}$ will be sensitive to even modest amounts of noise.

The effects of nearly common subchannel roots on CM receivers are quite similar. Since nearly common roots do not actually violate **(A1)**, the PBE conditions may still be satisfied, as shown by Fig. 2.9. They do excite a similar noise gain phenomenon, however, affecting CM receivers quite similarly to their Wiener counterparts (see Fig. 2.10). Note again that the better J_{cm} minima are in close proximity to the better Wiener

[12] Recall that with square invertible **H** and in the absence of noise the ZF and MMSE equalizers are identical. Thus the $*$ in Fig. 2.9 denote the locations of the ZF equalizers.

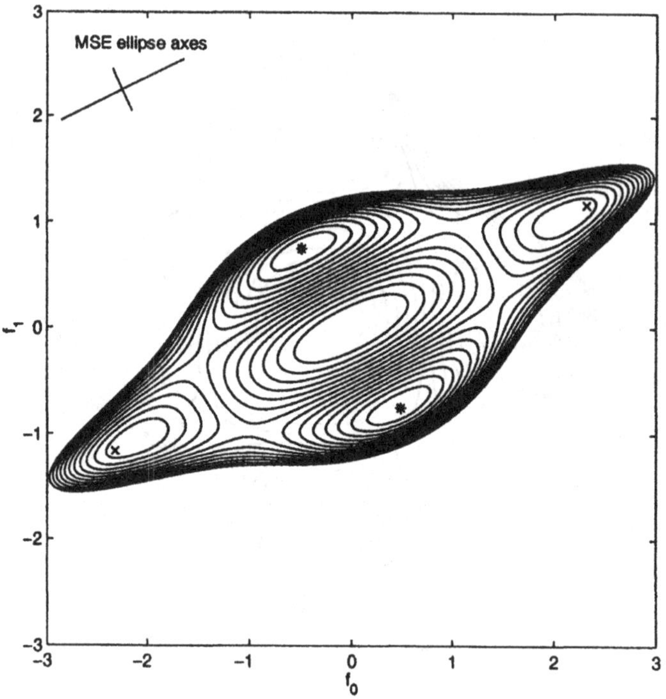

Figure 2.8 J_{cm} contours for channel $\mathbf{h}_b$ and 20-dB SNR in equalizer ($\mathbf{f}$) space.

solutions, offering evidence for Godard's conjecture. CM robustness to common subchannels has been formally addressed in Fijalkow et al. (1997).

Effect of Channel Undermodeling When the equalizer length is short enough to violate **(A1)**, the equalizer is, in general, not capable of perfect symbol recovery. This can be demonstrated by leaving our equalizer length at two taps but extending the $\mathbf{h}_b$ channel response to $(1.0, -0.5, 0.2, 0.3, -0.2, -0.15)^T = \mathbf{h}_d$. Now we have

$$\mathbf{H} = \begin{pmatrix} h_1 & h_3 & h_5 \\ h_0 & h_2 & h_4 \end{pmatrix}, \qquad \mathbf{f} = \begin{pmatrix} f_0 \\ f_1 \end{pmatrix}$$

The dimensions of $\mathbf{H}$ indicate that three distinct system delays are now possible, and thus there exist six Wiener receivers (after allowing $\pm$ system polarities). Looking at Fig. 2.11, however, the number of CM

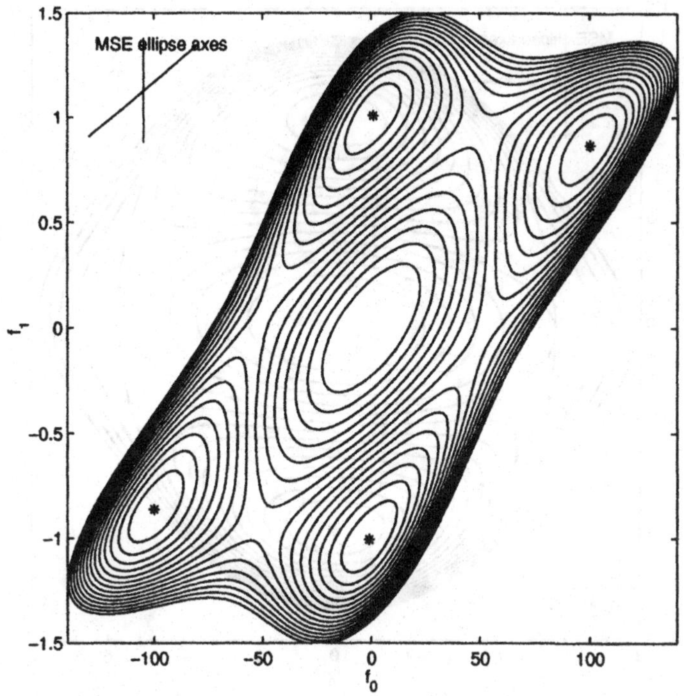

Figure 2.9 J_{cm} contours for nearly common subchannel-roots channel $\mathbf{h}_c$ and no noise in equalizer ($\mathbf{f}$) space. Note axis scaling.

receivers has not increased. It is important to note that the best CM minima are still near the best Wiener solutions, implying a certain degree of robustness to undermodeling and further evidence for Godard's conjecture. These robustness properties are discussed further in later subsection.

Effect of Source Correlation PBE condition **(A3)** specifies a white source sequence. To examine this condition, Fig. 2.12 shows the effect of a temporally correlated source on the CM cost surface using noiseless channel $\mathbf{h}_b$ and the 4-PAM periodic sequence $\{s_n\} = \{\ldots, 3, 1, -1, -3, \ldots\}$. In comparison to Fig. 2.6, we can observe a "twisting" of the cost surface that leads to a significant separation between the CM and Wiener receivers. More dramatic effects are possible in higher-dimensional systems, including a potential increase in the number of CM local minima (LeBlanc et al. 1998).

In the complex-valued case, correlation between the real and imaginary components of the source yields $E\{s_n^2\} \neq 0$, thus violating **(A3)**.

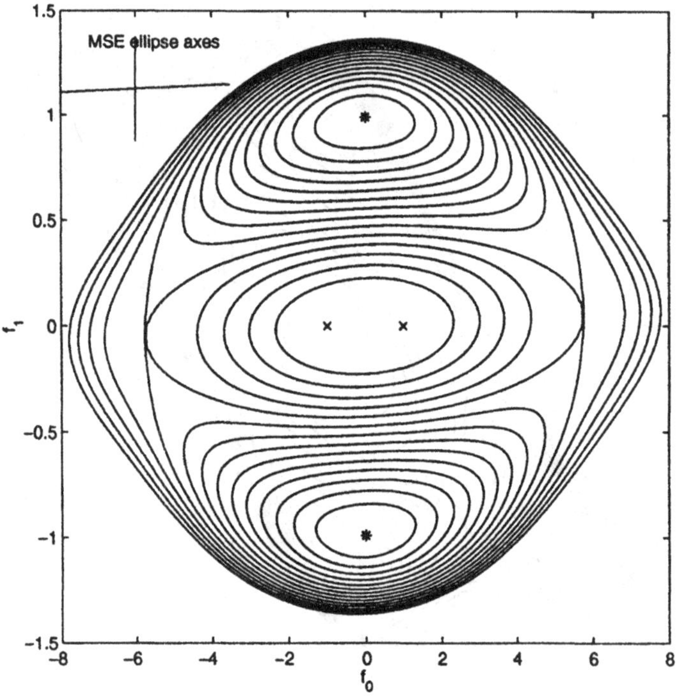

Figure 2.10 J_{cm} contours for nearly common subchannel-roots channel $\mathbf{h}_c$ and 20-dB SNR in equalizer ($\mathbf{f}$) space. Note axis scaling.

Axford et al. (1998) and Papadias (1997) discuss the presence of false CM minima under these conditions.

Effect of Source Kurtosis Condition **(A4)** states that the source distribution must be sub-Gaussian (i.e., $\kappa_s < \kappa_g$) for perfect blind equalizability. This requirement is satisfied by all uniformly distributed data constellations, such as M-PAM, M-PSK, and M-QAM. Even so, the value of the (sub-Gaussian) kurtosis still has an effect on the shape of the CM cost surface. Specifically, as the kurtosis (2.69) increases toward κ_g, the CM cost surface rises and all but the radial component of the gradient tends to vanish. Figure 2.13 shows the surface for a uniformly distributed 32-PAM source, for which $\kappa_s \approx 1.8$. (Compare to Fig. 2.5.) Note that the parameter-space positions of the CM minima are unaltered (as expected, since the PBE conditions are still satisfied). For more information on the effect of source kurtosis on the CM criterion, we refer interested readers to LeBlanc et al. (1998) and LeBlanc (1995).

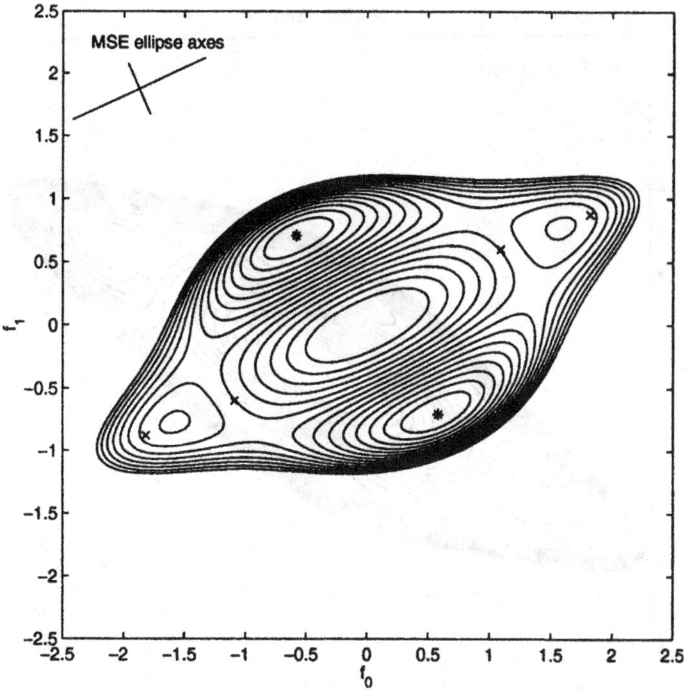

Figure 2.11 J_{cm} contours for undermodeled channel $\mathbf{h}_d$ and no noise in equalizer (**f**) space.

Properties of CM Stationary Points Recall that, when the PBE conditions are satisfied, all minima of the CM cost function are global minima and coincide with the ZF and Wiener receivers. In such a case, the CM receivers achieve perfect source recovery.

In addition to minima, the CM cost surface exhibits saddle points and one maximum. These extrema, commonly called stationary points, are described in Johnson and Anderson (1995) for binary phase-shift keying (BPSK) and a real-valued channel. With an i.i.d. source and in the absence of channel noise, the CM stationary points can be identified by having system responses $\mathbf{q}$ satisfying particular criteria. Specifically, a stationary point $\mathbf{q}_c$ satisfies the zero-gradient condition $\nabla_{\mathbf{f}} J_{cm} = 0$, where the equalizer-space gradient $\nabla_{\mathbf{f}} J_{cm}$ is given below as a function of $\mathbf{q}$ [under **(A2)** and **(A3)**]:

$$\nabla_{\mathbf{f}} J_{cm}(\mathbf{q}) = \sigma_s^4 \mathbf{H} \mathbf{\Delta_q} \mathbf{q} \tag{2.70}$$

The matrix $\mathbf{\Delta_q}$ is defined (in both the real- and complex-valued cases) as

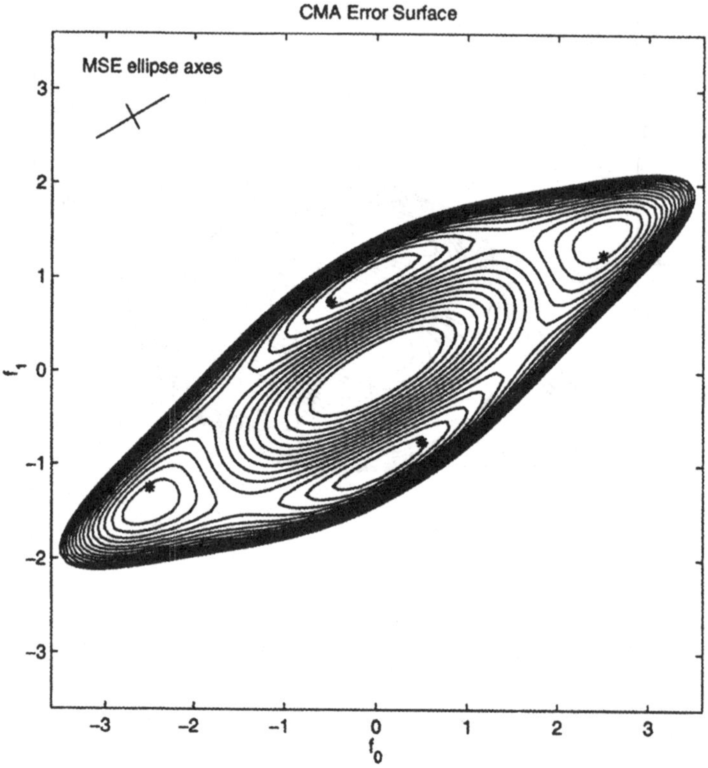

Figure 2.12 J_{cm} contours for channel $\mathbf{h}_b$ and periodic source in equalizer ($\mathbf{f}$) space.

follows:

$$\boldsymbol{\Delta}_{\mathbf{q}} \triangleq (\kappa_g \|\mathbf{q}\|_2^2 - \kappa_s)\mathbf{I} - (\kappa_g - \kappa_s)\,\mathrm{diag}(\mathbf{q}\mathbf{q}^H) \tag{2.71}$$

where κ_g is the kurtosis of a Gaussian source. The matrix operation diag$(\cdot)$ retains the diagonal entries of its argument while replacing the other entries with zeros.

The stability of a CM stationary point (i.e., whether or not it is a local minima) is given by the positive-semidefiniteness of its equalizer-space Hessian, denoted by $\mathscr{H}_{\mathbf{f}} J_{cm} \geq 0$. The Hessian is given below, for the case of real-valued quantities,[13] as a function of $\mathbf{q}$ [again, under **(A2)** and

[13] The complex-valued case requires calculating the partial derivatives of J_{cm} with respect to the real part and imaginary part of $\mathbf{f}$ separately. For example, the Hessian could be taken with respect to the real-valued vector $\bar{\mathbf{f}} = (\mathrm{Re}\,\mathbf{f}^T, \mathrm{Im}\,\mathbf{f}^T)^T$.

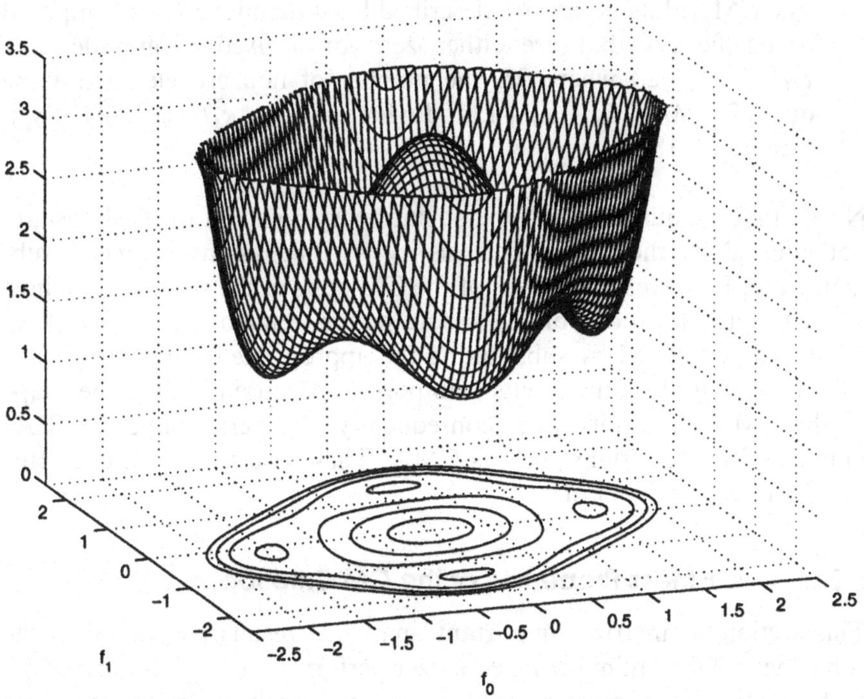

Figure 2.13 $J_{\rm cm}$ for non-CM source ($\kappa_s \approx 1.8$) and channel $\mathbf{h}_a$ in equalizer ($\mathbf{f}$) space.

(A3)]:

$$\mathscr{H}_{\mathbf{f}} J_{\rm cm} = \sigma_s^4 \mathbf{H} \boldsymbol{\Psi}_{\mathbf{q}} \mathbf{H}^T \tag{2.72}$$

The matrix $\boldsymbol{\Psi}_{\mathbf{q}}$ is defined as follows:

$$\boldsymbol{\Psi}_{\mathbf{q}} \triangleq (\kappa_g \|\mathbf{q}\|_2^2 - \kappa_s)\mathbf{I} + 2\kappa_g \mathbf{q}\mathbf{q}^T - 3(\kappa_g - \kappa_s)\,\mathrm{diag}(\mathbf{q}\mathbf{q}^T) \tag{2.73}$$

In the case that $\mathbf{H}$ is full column rank, the requirements for $\mathbf{q}$ to be a CM minimum reduce to $\boldsymbol{\Delta}_{\mathbf{q}}\mathbf{q} = 0$ and $\boldsymbol{\Psi}_{\mathbf{q}} \geq 0$.

In summary, it can be shown [see, e.g., Fijalkow et al. (1994) for the real-valued case] that, under the PBE conditions,

- All CM minima correspond to ZF system responses, that is, $\mathbf{q}_c = \pm \mathbf{e}_\delta$ for $0 \geq \delta \geq L_f + L_h$.
- The only CM maximum occurs at the origin, that is, $\mathbf{q} = 0$.

- The CM saddle points are described by system responses composed of coefficients that are either zero or a fixed magnitude: $\mathbf{q} = \tau_M \sum_{i=0}^{M-1} (\pm \mathbf{e}_i)$ where M, the number of nonzero elements in $\mathbf{q}$, obeys $2 \leq M \leq L_f + L_h + 1$, and where $\tau_M = \sqrt{\kappa_s / (\kappa_g (M-1) + \kappa_s)}$. (See Fig. 2.18.)

Notice that the number and type of stationary points of the CM cost are not affected by the source kurtosis κ_s (as long as the source is sub-Gaussian). However, the values of the CM costs at the stationary points, as well as the locations of the saddle points, remain a function of κ_s. (This can be verified by substituting the appropriate $\mathbf{q}$ into Eq. (2.68).) The relocation of saddle points is important because it affects the shape of the CM cost surface and, consequently, the performance of SGD minimization algorithms such as CMA. These issues are discussed further in an later subsection.

2.3.2 Robustness Properties of the CM Criterion

This section summarizes important analytical results pertaining to the robustness of CM-minimizing equalizer performance to violations of the PBE conditions. Though we will focus on the robustness to noise and channel undermodeling, we note that evidence of CM robustness to nearly common subchannel roots appears in Fijalkow et al. (1997), and evidence of robustness to correlated sources appears in LeBlanc et al. (1998).

The first two subsections below describe performance approximations obtained using perturbation techniques. In general, such techniques measure the effect of local deviations from a reference setting (e.g., a ZF or MMSE receiver) and are generally valid for small deviations about this reference. Later in the section, we will study the effect of noise on the CM criterion using a technique that does not rely on perturbational approximations, but instead uses a geometric approach to generate tight bounds on performance.

Perturbation in Noise Power Noise is unconditionally present in practical environments, thus violating **(A2)**. In such cases, both the achieved system delay and the distance between subchannel roots may significantly impact the performance of a CM receiver (Fijalkow 1996). A small amount of noise can be viewed as a perturbation of the (noise-free) parameters of a given equalizer design. For example, noise-perturbation analysis has been considered around noise-free CM receivers in Fijalkow et al. (1997) and Fijalkow et al. (1995), ZF receivers in Li et al. (1996), and Wiener receivers in Zeng et al. (1996a,b). This method is justified by the fact that the CM-cost is a smooth (polynomial) function of the equalizer parameters, and assumes that a CM receiver is indeed in the neighborhood of the ZF or MMSE receiver around which the perturbation analysis is performed. Such an assumption may not always be valid, as shown in Chung and LeBlanc (1998) and Gu and Tong (1998).

Following the approach in Fijalkow et al. (1997), the CM-cost (for a unit-variance source and Gaussian noise) can be written in terms of the noise-free cost plus a term proportional to the noise power σ_w^2:

$$J_{\mathrm{cm}} = \tilde{J}_{\mathrm{cm}} + \frac{\sigma_w^2}{4}\|\mathbf{f}\|_2^2(2(\kappa_g\|\mathbf{H}^T\mathbf{f}\|_2^2 - \kappa_s) + \kappa_g\sigma_w^2\|\mathbf{f}\|_2^2) \tag{2.74}$$

where $\tilde{J}_{\mathrm{cm}}$ is the following noise-free CM cost function:

$$\tilde{J}_{\mathrm{cm}} \triangleq \frac{1}{4}\mathrm{E}\left\{\left(|\mathbf{f}^T\mathbf{H}\mathbf{s}(n)|^2 - \frac{\mathrm{E}\{|s_n|^4\}}{\sigma_s^2}\right)^2\right\} \tag{2.75}$$

The equalizer minimizing J_{cm} must therefore strike a balance between the noise enhancement appearing as the second term of Eq. (2.74) and the CM penalty imposed by $\tilde{J}_{\mathrm{cm}}$. The behavior of CM receivers in the presence of noise can then be analyzed using a Taylor series expansion of the CM cost around the noise-free CM receiver. When **(A1)** is satisfied, Fijalkow et al. (1997) use a first-order approximation of **q**-space minima relocation as a function of noise power to yield the following MSE approximation:

$$J_{\mathrm{m}}^{(\delta)}(\mathbf{f}_{\mathrm{c}}^{(\delta)}) \approx \sigma_w^2\mathbf{e}_\delta^t(\mathbf{H}^t\mathbf{H})^{-1}\mathbf{e}_\delta + o(\sigma_w^2)$$

We note that the preceding expression is equivalent to a first-order noise-power expansion of MSE about the Wiener receiver. Under lack of disparity [i.e., the presence of common subchannel roots—violating **(A1)**] the noise-free CM receivers are difficult to express analytically. However, when the equalizer is long enough to mitigate the baud-spaced channel component lacking disparity [see Eq. (2.24)], sufficiently long CM receivers have been shown to yield MSE similar to that of the Wiener receivers when operating in the presence of channel noise (Fijalkow et al. 1997).

Li et al. (1996) study the effect of noise on the baud-spaced CM receiver relative to the (IIR) ZF receiver. Similar to the techniques described earlier, they approximate the resulting MSE using a Taylor series expansion. In the presence of noise, the CM receivers are shown to yield lower MSE than the ZF receivers. Since the ZF receivers are susceptible to noise gain, this result should not come as a surprise. Recall that the presence of nearly common subchannel roots can make the performance of ZF receivers very sensitive to noise.

Leveraging Godard's conjecture concerning the proximity of CM receivers to Wiener receivers, Zeng et al. (1996a,b) perform a noise-perturbation analysis around the Wiener receiver. Under satisfaction of the remaining PBE conditions, this approach shows that the CM receiver is collinear with the Wiener receiver up to $O(\mathscr{E}_{\mathrm{m}}^3)$, where we have been using $\mathscr{E}_{\mathrm{m}}$ to denote the MSE of the Wiener receiver.

Perturbation in Equalizer Length Endres et al. (1997a,b) study the robustness properties of the CM criterion (and of CMA) to the sub-optimal but realistic situation in which the number of FSE coefficients is less than that needed to perfectly cancel ISI, thus violating **(A1)**. The main conclusions of this work include evidence that "small" under-modeled channel coefficients produce only a mild deformation of the CM cost surface, and that Wiener receivers corresponding to system delays of better MSE performance have CM minima in closer proximity than those corresponding to system delays of worse performance. Endres's work also results in design guidelines for the choice of equalizer length, the subject of Section 2.4.

Endres presents an algebraic analysis describing the deformation of the CM cost surface from the case of PBE satisfaction as a result of (1) channel coefficients outside the time span of the FSE, and (2) equalizer coefficients lost in truncation, both violating **(A1)**. Using microwave-channel models from the signal-processing information base (SPIB)[14] database, it is shown that when the undermodeled channel coefficients are "small," the change in the CM cost is "small" and approximately equal to the scaled change in the MSE cost, suggesting that the Wiener minima and CM minima stay in a tight neighborhood.

As a means of studying the relationship between undermodeled CM cost and achieved system delay, Endres (1997a,b) construct a second-order Taylor series approximation of the binary CM cost function about the Wiener solution. Since this approximation is quadratic in the FSE coefficients, a closed-form expression can be obtained for approximations of the CM minima corresponding to various system delays. A

[14] The SPIB database resides at `http://spib.rice.edu/spib/select_comm.html`.

measure of the proximity of CM and Wiener minima is derived, suggesting that the Wiener receivers corresponding to system delays of good MSE performance have CM receivers in closer proximity than do the Wiener receivers corresponding to system delays of worse MSE performance.

2.3.3 Exact Analysis of the CM Criterion

Although both Godard (1980) and Treichler and Agee (1983) suggested that CMA offers approximately Wiener-like MSE performance, it was Foschini (1985) who first established the equivalence between the CM and Wiener receivers under the perfect conditions outlined earlier. Ding et al. (1991) first established the potential for CM local minima under violation of the SPE condition **(A1)**. Such loss of SPE prevents, in general, the equivalence between CM and Wiener receivers.

The presence of noise, violating **(A2)**, is another factor preventing the theoretical equivalence between the CM and Weiner receivers. In fact, most CM analysis is made quite difficult by the presence of noise. For example, the task of characterizing the full set of CM-cost stationary points—let alone distinguishing that are local minima—becomes far more complicated than in the noiseless SPE case. Although the perturbation analyses previously described can be used to assess the MSE performance of CM receivers (and are reasonably accurate at high SNR), such approaches rely on *approximations* of the CM cost function and, as such, have two main drawbacks. First, they presume the existence of a CM receiver in a neighborhood of the reference (e.g., the ZF or Wiener receiver), which may not be the case. Second, they make it difficult to characterize the accuracy of the analysis given a particular SNR.

We describe here an alternative approach (Zeng et al. 1998) to the analysis of the CM criterion that does not involve an approximation of the CM cost function. This approach, commonly referred to as the "exact CM analysis," yields the following results:

1. A signal-space property and power constraints that apply to all CM receivers;
2. An analytical expression that can be used to verify the existence of a CM local minimum in a particular neighborhood of a Wiener receiver;
3. An analytical description of the regions that may contain CM local minima in the neighborhoods of Wiener receivers; and
4. Upper and lower bounds on the MSE of these CM receivers.

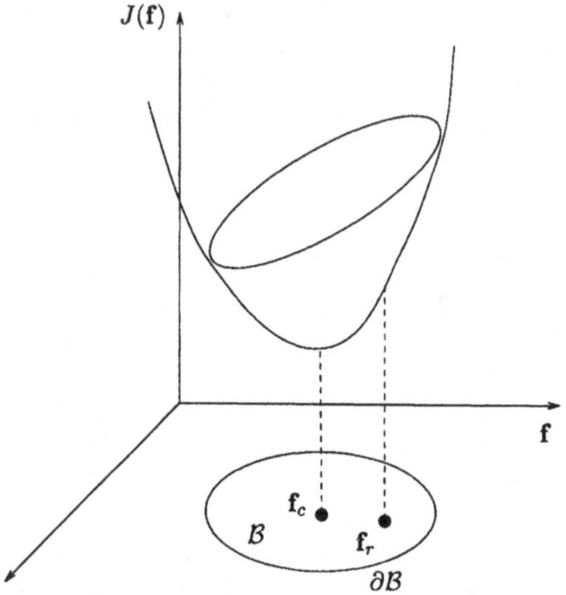

Figure 2.14 Illustration of the main idea underlying the "exact analysis."

Unlike techniques based on equilibrium-point analysis and CM cost-surface local curvature, this approach to analyzing FIR CM receivers is geometrical and is based on the Weierstrass maximum theorem (Luenberger 1990, 40). As illustrated in Fig. 2.14, we attempt to find a compact region of equalizer parameter space, called $\mathscr{B}$, with boundary $\partial\mathscr{B}$ and interior reference $\mathbf{f}_r$, such that the CM cost J_{cm} everywhere on the boundary is greater than the CM cost at the reference. As a consequence of the Weierstrass theorem, the continuity of the CM cost function implies that at least one CM local minimum exists in $\mathscr{B}$.

Critical in this analysis is the selection of the shape of $\mathscr{B}$, the location of $\mathscr{B}$, and the location of the reference $\mathbf{f}_r$. In defining the region $\mathscr{B}$, the first goal is for it to be as small as possible. This should lead to more accurate descriptions of the local minimum and its MSE performance. The second goal is to relate the region $\mathscr{B}$ to the location of the neighboring Wiener receiver. In addition to the description of $\mathscr{B}$ in the parameter space, we give a description in the Hilbert space of the observations that provides important physical interpretations.

We emphasize that these results apply only to CM local minima sufficiently near the corresponding Wiener solutions. Therefore, if CM local minima exist within the regions defined by our approach, then one of these minima must be the globally MSE-optimal CM receiver.

The Signal-Space Property of CM Receivers As pointed out in Ericson (1971), a receiver designed by any "reasonable" criterion of goodness includes a matched filter as its front end. Recalling that system model (2.16) formulates the time-n received vector as $\mathbf{r}(n) = \mathbf{H}\mathbf{s}(n) + \mathbf{w}(n)$, the presence of a matched filter is equivalent to a receiver having the *signal-space property*: the receiver is a linear combination of the columns of the channel matrix $\mathbf{H}$. The signal-space property is important for several reasons. First, receivers with the signal-space property will filter out all noise orthogonal to the signal space. Second, under the SPE condition *and* the signal-space property, the analysis of the receiver parameter $\mathbf{f}$ is equivalent to the analysis of system-response parameter $\mathbf{q}$. The equivalence between the equalizer and system spaces considerably simplifies the analysis that follows.

Noting that the output power can be written $E\{|y_n|^2\} = \mathbf{f}^T E\{\mathbf{r}(n)\mathbf{r}^T(n)\}\mathbf{f} = \mathbf{f}^T\mathbf{R}_{r,r}\mathbf{f} = \|\mathbf{f}\|^2_{\mathbf{R}_{r,r}}$, we state what is known as the *CM power constraint*:

Theorem 2.1. All local minima of the CM cost function are in the signal subspace $\mathscr{Y}_{\mathbf{H}}$. (That is, for any CM local minimum $\mathbf{f}_c$, there exists a vector $\mathbf{v}$ such that $\mathbf{f}_c = \mathbf{H}\mathbf{v}$.) Furthermore, when $\sigma_w^2 > 0$, the output power of any CM receiver $\mathbf{f}_c$ satisfies

$$\frac{\kappa_s}{3} < \|\mathbf{f}_c\|^2_{\mathbf{R}_{r,r}} < 1 \tag{2.76}$$

Proof. See Zeng (1998).

For BPSK transmission in the absence of channel noise, the CM power constraint, Eq. (2.76) first appeared in Johnson and Anderson (1995). Geometrically, it implies that *all* CM receivers lie in an elliptical "shell" in parameter space, as illustrated by Fig. 2.15 for the two-parameter case.

It is interesting to compare the output power of a CM receiver with that of the ZF and Wiener receivers. For the ZF receiver corresponding to system delay δ, we have (assuming $\sigma_s^2 = 1$ and $\sigma_w^2 > 0$)

$$\|\mathbf{f}_z^{(\delta)}\|^2_{\mathbf{R}_{r,r}} = \mathbf{e}_\delta^T\mathbf{H}^\dagger(\mathbf{H}\mathbf{H}^T + \sigma_w^2\mathbf{I})(\mathbf{H}^T)^\dagger\mathbf{e}_\delta = 1 + \sigma_w^2\|\mathbf{f}_z^{(\delta)}\|_2^2 > 1$$

Contrast this to the Wiener receiver, for which it is known that

$$\|\mathbf{f}_m^{(\delta)}\|^2_{\mathbf{R}_{r,r}} < 1$$

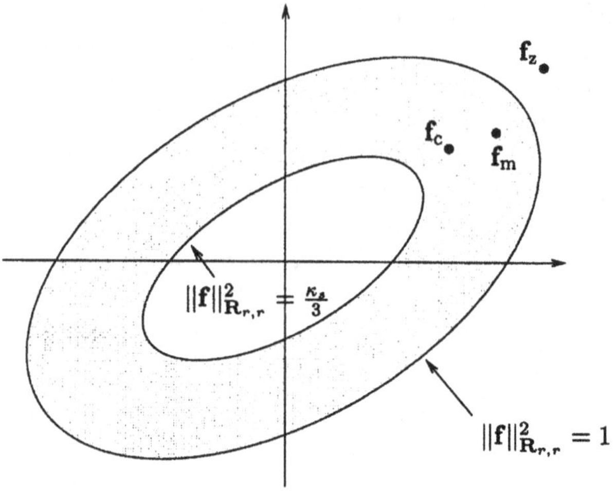

Figure 2.15 Region of CM local minima.

Therefore, the output powers of Wiener receivers are always less than 1, whereas the output powers of ZF receivers are greater than 1. As SNR decreases to zero, the output power of Wiener receivers approaches zero and the output power of ZF receivers approaches infinity, but the output power of CM receivers stays between $\kappa_s/3$ and 1. This condition, particularly the lower bound, is useful in determining if a CM local minimum exists near the Wiener receiver.

In the sequel, we assume that the SPE condition **(A1)** is satisfied (i.e., $\mathbf{H}$ is full column rank). Thanks to the signal-space property, portions of the following analysis are performed in the system parameter (i.e., $\mathbf{q}$) space.

Location of CM Receivers Our estimates of the CM receiver's location and achieved MSE are obtained by first specifying a neighborhood $\mathcal{B}$ near the Wiener receiver and then comparing the CM cost on the boundary $\partial\mathcal{B}$ with that of a reference $\mathbf{q}_{\mathrm{r}}$ contained in $\mathcal{B}$.

For the remainder of the section we assume that the source is BPSK, implying that $\kappa_s = \sigma_s^2 = 1$.

The Neighborhood The neighborhood has different but equivalent definitions in the equalizer parameter space and the equalizer output space. To keep the notation simple, we consider a receiver $\mathbf{f}$ that estimates the first symbol s_0 of a transmitted source vector $\mathbf{s}$ given the noisy observation $\mathbf{r} = \mathbf{Hs} + \mathbf{w}$. In this context, the receiver gain θ and the conditionally unbiased MSE (UMSE) are defined as follows:

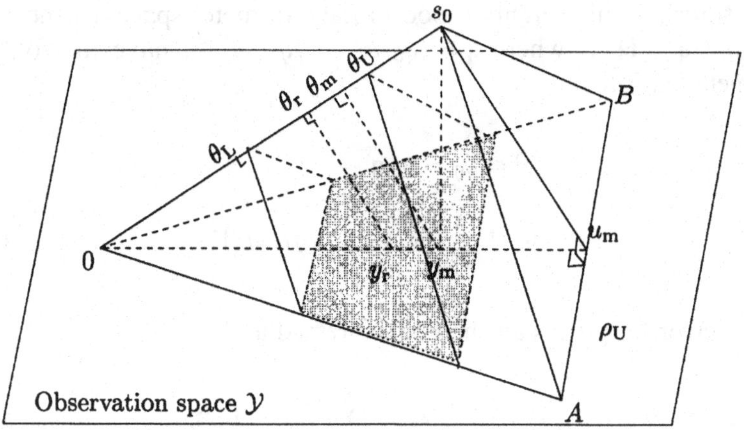

Figure 2.16 The region $\mathscr{B}$ in the Hilbert space of the observations.

$$\theta \triangleq \mathbf{f}^T\mathbf{H}\mathbf{e}_0, \qquad \text{UMSE} \triangleq \mathrm{E}\left\{\left|\frac{1}{\theta}\mathbf{f}^T\mathbf{r} - s_0\right|^2\right\} \tag{2.77}$$

Note that $(1/\theta)\mathbf{f}^T\mathbf{r}$ is a conditionally unbiased estimate of s_0 in the sense that

$$\mathrm{E}\left\{\frac{1}{\theta}\mathbf{f}^T\mathbf{r}|s_0\right\} = s_0 \tag{2.78}$$

Shown in Fig. 2.16 is the geometry associated with the linear estimation of s_0 from $\mathbf{r}$. The output of any linear estimator must lie in the plane $\mathscr{Y}$ spanned by the components of $\mathbf{r}$. The output y_m of the Wiener receiver is obtained by projecting s_0 onto $\mathscr{Y}$. If we scale y_m to u_m such that the projection of u_m onto the direction of s_0 is s_0 itself, we obtain the so-called (conditionally) unbiased minimum mean-square-error (U-MMSE) estimate of s_0. Indeed, u_m is conditionally unbiased since $E\{u_m|s_0\} = s_0$. Furthermore, it is easy to see from Fig. 2.16 that u_m has the shortest distance to s_0 (and hence the minimum MSE) among all conditionally unbiased estimates. Note that the output of a conditionally unbiased estimator must lie on the line $\overline{AB}$ in the figure.

Shown in the shaded area of Fig. 2.16 is a neighborhood of estimates whose receiver gain (obtained by projecting the estimate in the direction of s_0) is bounded in the interval (θ_L, θ_U), and whose corresponding conditionally unbiased estimates of s_0 have mean-square error at most ρ_U^2 greater than MSE(u_m). In other words, the estimates in the shaded region have extra UMSE upper bounded by ρ_U^2.

To translate this neighborhood to the parameter space, let the system response $\mathbf{q} = \mathbf{H}^T\mathbf{f}$, where $\mathbf{q} = (q_0, q_1, \ldots, q_{L_f+L_h})^T$, have the following parameterization:

$$\theta \triangleq q_0 = \mathbf{e}_0^T \mathbf{q} \tag{2.79}$$

$$\mathbf{q}_\mathrm{I} \triangleq \frac{1}{q_0}(q_0, q_1, \ldots, q_{L_f+L_h})^T \tag{2.80}$$

The receiver output y can then be expressed as

$$y = \underbrace{\theta}_{\text{gain}} \cdot s_0 + \underbrace{\sum_{i \neq 0} q_i s_i}_{\text{interference}} + \underbrace{\mathbf{f}^T \mathbf{w}}_{\text{noise}} \tag{2.81}$$

where θ is the receiver gain, or bias. Scaling y by $1/\theta$, we have the (conditionally) unbiased estimate of s_0:

$$u \triangleq \frac{y}{\theta} = s_0 + \underbrace{\frac{1}{\theta}\left(\sum_{i \neq 0} q_i s_i + \mathbf{f}^T \mathbf{w}\right)}_{\text{equivalent noise}} \tag{2.82}$$

Therefore, the receiver gain and UMSE of $\mathbf{q}$ is given by θ and $\mathrm{MSE}(u)$, respectively. Hence, the shaded neighborhood in Fig. 2.16 can be described as

$$\{\mathbf{q} : \theta_\mathrm{L} < \theta < \theta_\mathrm{U}, \mathrm{MSE}(u) - \mathrm{MSE}(u_\mathrm{m}) < \rho_\mathrm{U}^2\} \tag{2.83}$$

In this definition, $\theta_\mathrm{L}(\theta_\mathrm{U})$ specifies the lower (upper) bound on CM receiver gain, while ρ_U^2 specifies the upper bound on extra UMSE.

Although the neighborhood defined in (2.83) is related to specific characteristics of a receiver (i.e., UMSE and bias), its relationship to the receiver coefficients ($\mathbf{q}$) is not given explicitly. To locate the CM receiver using this neighborhood, it is necessary to translate the neighborhood of (2.83) to one that is specified in terms of the system parameter space. Given the Wiener receiver $\mathbf{q}_\mathrm{m} = \theta_\mathrm{m}\begin{pmatrix} 1 \\ \mathbf{q}_\mathrm{mI} \end{pmatrix}$, it can be shown (Zeng et al. 1998) that an equivalent neighborhood, illustrated in the parameter space by Fig. 2.17, is given by

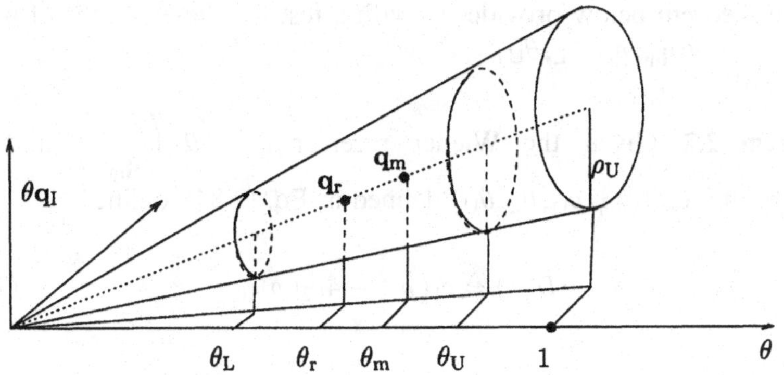

Figure 2.17 A cone-type region: $\mathscr{B}(\mathbf{q}_m, \rho_U, \theta_L, \theta_U)$.

$$\mathscr{B}(\mathbf{q}_{\rm m}, \rho_{\rm U}, \theta_{\rm L}, \theta_{\rm U}) \triangleq \{\mathbf{q} : \theta_{\rm L} < \theta < \theta_{\rm U}, \|\mathbf{q}_{\rm I} - \mathbf{q}_{\rm mI}\|_{\rm C} < \rho_{\rm U}\} \tag{2.84}$$

$$= \{\mathbf{q} : \theta_{\rm L} < \theta < \theta_{\rm U}, \quad \mathrm{MSE}(u) - \mathrm{MSE}(u_{\rm m}) < \rho_{\rm U}^2\} \tag{2.85}$$

where matrix **C** is a matrix formed by removing the first row and column from $(\mathbf{I} + \sigma_w^2 \mathbf{H}^\dagger (\mathbf{H}^\dagger)^T)$.

The Reference q_r In relating the CM receiver to its Wiener counterpart, we choose, in the direction of the Wiener receiver $\mathbf{q}_{\rm m} = \theta_{\rm m}\begin{pmatrix} 1 \\ \mathbf{q}_{\rm mI} \end{pmatrix}$, the reference point

$$\mathbf{q}_r \triangleq \theta_{\rm r}\begin{pmatrix} 1 \\ \mathbf{q}_{m\rm I} \end{pmatrix} \tag{2.86}$$

with the minimum CM cost. In other words, $\theta_{\rm r}$ is chosen to minimize the CM cost of $\mathbf{q} = \theta\begin{pmatrix} 1 \\ \mathbf{q}_{\rm mI} \end{pmatrix}$ with respect to θ, which can be shown to obey

$$\theta_{\rm r}^2 = \frac{\theta_{\rm m}}{3 - 2\theta_{\rm m}^2 - 2\theta_{\rm m}^2\|\mathbf{q}_{\rm mI}\|_4^4} \tag{2.87}$$

In analyzing CM local minima in the neighborhood of Wiener receivers, the role of $\mathbf{q}_{\rm r}$ turns out to be more than only technical: it can be shown that $\mathbf{q}_{\rm r}$ is a very good approximation of the CM receiver.

The theorem below provides us with a test for the existence of a CM receiver in $\mathscr{B}(\mathbf{q}_{\mathrm{m}}, \rho_{\mathrm{U}}, \theta_{\mathrm{L}}, \theta_{\mathrm{U}})$.

Theorem 2.2. Given the Wiener receiver $\mathbf{q}_{\mathrm{m}} = \theta_{\mathrm{m}} \begin{pmatrix} 1 \\ \mathbf{q}_{\mathrm{mI}} \end{pmatrix}$ and the neighborhood $\mathscr{B}(\mathbf{q}_{\mathrm{m}}, \rho_{\mathrm{U}}, \theta_{\mathrm{L}}, \theta_{\mathrm{U}})$ defined in Eq. (2.84), define

$$D(\rho) \triangleq c_1(\rho)^2 - 4c_2(\rho)c_0 \tag{2.88}$$

where

$$c_0 = \frac{1}{3 - 2\theta_{\mathrm{m}}^2 - 2\theta_{\mathrm{m}}^2 \|\mathbf{q}_{\mathrm{mI}}\|_4^4} \tag{2.89}$$

$$c_1(\rho) = -2\left(\rho^2 + \frac{1}{\theta_{\mathrm{m}}}\right) \tag{2.90}$$

$$c_2(\rho) = 3\left(\rho^2 + \frac{1}{\theta_{\mathrm{m}}}\right)^2 - 2(1 + (\rho + \|\mathbf{q}_{\mathrm{mI}}\|_4)^4) \tag{2.91}$$

Under the conditions that MMSE $< \frac{1}{3}$ and $D\,(\|\mathbf{q}_{\mathrm{mI}}\|_2) < 0$, a CM local minimum exists in the neighborhood $\mathscr{B}_p(\mathbf{q}_{\mathrm{m}}, \rho_{\mathrm{U}}^*, \theta_{\mathrm{L}}^*, \theta_{\mathrm{U}}^*)$. Here, ρ_{U}^* is the smallest positive root of $D(\rho)$, and

$$\theta_{\mathrm{L}}^* = \min_{0 \le \rho \le \rho_{\mathrm{U}}^*} \sqrt{\frac{-c_1(\rho) - \sqrt{c_1(\rho)^2 - 4c_2(\rho)c_0}}{2c_2(\rho)}} \tag{2.92}$$

$$\theta_{\mathrm{U}}^* = \max_{0 \le \rho \le \rho_{\mathrm{U}}^*} \sqrt{\frac{-c_1(\rho) + \sqrt{c_1(\rho)^2 - 4c_2(\rho)c_0}}{2c_2(\rho)}} \tag{2.93}$$

Furthermore, as the noise power decreases,

$$\lim_{\sigma_w \to 0} \mathscr{B}_p(\mathbf{q}_{\mathrm{m}}, \rho_{\mathrm{U}}^*, \theta_{\mathrm{L}}^*, \theta_{\mathrm{U}}^*) = \lim_{\sigma_w \to 0} \mathbf{q}_{\mathrm{c}} = \lim_{\sigma_w \to 0} \mathbf{q}_{\mathrm{m}} = \mathbf{q}_{\mathrm{z}} \tag{2.94}$$

Proof. See Zeng et al. (1998).

Remarks on Theorem 2.2. Some comments on the implications of Theorem 2.2 follow:

- Given $\mathbf{q}_{\rm m} = \theta_{\rm m}\begin{pmatrix} 1 \\ \mathbf{q}_{\rm mI} \end{pmatrix}$, this theorem can be used (1) to detect the existence of a CM local minimum by checking the sufficient condition $D(\|\mathbf{q}_{\rm mI}\|_2) < 0$, and (2) to determine the region $\mathscr{B}_p(\mathbf{q}_{\rm m}, \rho_{\rm U}^*, \theta_{\rm L}^*, \theta_{\rm U}^*)$.
- Equation (2.94) states that the region $\mathscr{B}_p(\mathbf{q}_{\rm m}, \rho_{\rm U}^*, \theta_{\rm L}^*, \theta_{\rm U}^*)$ shrinks to the Wiener/ZF receiver as noise vanishes, which is intuitively satisfying.

MSE of CM Receivers Theorem 2.2 also enables us to give performance bounds on the MSE of CM receivers.

Theorem 2.3. Given the Wiener receiver $\mathbf{q}_{\rm m} = \theta_{\rm m}\begin{pmatrix} 1 \\ \mathbf{q}_{\rm mI} \end{pmatrix}$, suppose that the MMSE is less than $\frac{1}{3}$, the condition in Theorem 2.2 is satisfied, and $\mathbf{q}_{\rm c} \in \mathscr{B}_p(\mathbf{q}_{\rm m}, \rho_{\rm U}^*, \theta_{\rm L}^*, \theta_{\rm U}^*)$ is a CM minimum. Then, letting $\Delta\mathscr{E}$ be the extra MSE of $\mathbf{q}_{\rm c}$ defined by $\Delta\mathscr{E} \triangleq J_{\rm m}((\mathbf{H}^T)^\dagger \mathbf{q}_{\rm c}) - J_{\rm m}((\mathbf{H}^T)^\dagger \mathbf{q}_{\rm m})$,

$$\underbrace{\frac{(\theta_{\rm U}^* - \theta_{\rm m})^2}{\theta_{\rm m}}}_{\Delta\mathscr{E}_{\rm L}} \leq \Delta\mathscr{E} \leq \underbrace{\frac{(\theta_{\rm L}^* - \theta_{\rm m})^2}{\theta_{\rm m}} + (\theta_{\rm U}^* \rho_{\rm U}^*)^2}_{\Delta\mathscr{E}_{\rm U}} \tag{2.95}$$

Proof. See Zeng et al. (1998).

The bounds in Eq. (2.95) involve the computation of $\theta_{\rm L}^*, \theta_{\rm U}^*$, and $\rho_{\rm U}^*$. Note that if the $\mathscr{B}$ is small in size, then the MSE of the CM receiver can be approximated by the MSE of the reference, and the location of the CM receiver $\mathbf{q}_{\rm c}$ can by approximated by the reference location $\mathbf{q}_{\rm r}$ (which, we recall, is a scaled version of the Wiener receiver).

2.3.4 Behavior of the Constant Modulus Algorithm

Having established the robustness of the CM criterion—that is, the close relationship between CM and MSE minima under violation of the perfect conditions **(A1)** to **(A4)**—we now turn to examination of the behavior of the CMA, which accomplishes a SGD of the CM cost surface. CMA can be described by the update equation

$$\begin{aligned} \mathbf{f}(n+1) &= \mathbf{f}(n) + \mu \mathbf{f}^T(n)\mathbf{r}(n)(\gamma - |\mathbf{f}^T(n)\mathbf{r}(n)|^2)\mathbf{r}^*(n) \\ &= \mathbf{f}(n) + \mu \underbrace{y_n(\gamma - |y_n|^2)}_{\text{CM error function}} \mathbf{r}^*(n) \end{aligned} \tag{2.96}$$

Recall that γ is the dispersion constant typically chosen equal to $(\mathrm{E}\{|s_n|^4\}/\sigma_s^2)$, $\mathbf{f}(n)$ is the equalizer impulse-response vector at time n, and μ is a small step-size. Though Eq. (2.96) presents the complex-valued CMA update equation, we focus the remainder of the section on strictly real-valued quantities. Averaging theory (Benveniste et al. 1990; Ljung and Soderstrom 1983) predicts that an SGD algorithm such as CMA converges almost surely and in mean to the local minima of its cost surface (i.e., the CM receivers).

The performance of CMA can be categorized by its transient and steady-state behaviors. The multimodal nature of the CM cost surface makes analysis of CMA transient behavior (e.g., the rate of convergence to steady-state operation) more difficult than with unimodal linear-equalizer update algorithms such as LMS. CMA's steady-state behavior is characterized by parameter "jitter" around the (local) CM receiver, resulting in nonzero excess CM-cost. In certain cases, the excess CM-cost can be translated directly into excess MSE, a popular measure of adaptive equalizer performance. These topics are discussed in more detail in the following subsections.

Transient Behavior: Convergence Rate From a given initialization, CMA updates the equalizer by descending the CM cost surface according to estimates of the direction of steepest descent. Due to the multimodality of J_{cm}, the convergence time (viewed as the number of iterations to reach "steady-state" behavior) is greatly affected by equalizer initialization. Consequently, very few studies exist on CMA convergence rate, Larimore and Treichler (1983), Touzni and Fijalkow (1996), and Lambotharan et al. (1997) being exceptions.

To analyze CMA convergence, we consider the average CMA trajectories, or in other words, the trajectories following an (exact) gradient descent of the CM cost surface in equalizer space. For a very small step-size μ, an SGD algorithm of the form

$$\mathbf{f}(n+1) = \mathbf{f}(n) + \mu F(\mathbf{f}(n), \mathbf{r}(n))$$

closely follows the average system [see Benveniste et al. (1990) and Ljung and Soderstrom (1983)] described by the ordinary differential equation (ODE):

$$\frac{d\mathbf{f}(t)}{dt} = \bar{F}(\mathbf{f})$$

where $\bar{F}(\mathbf{f}) = \mathrm{E}\{F(\mathbf{f}, \mathbf{r}(n))\}$ represents the mean update term of the algorithm and where t is a continuous variable proportional to n.

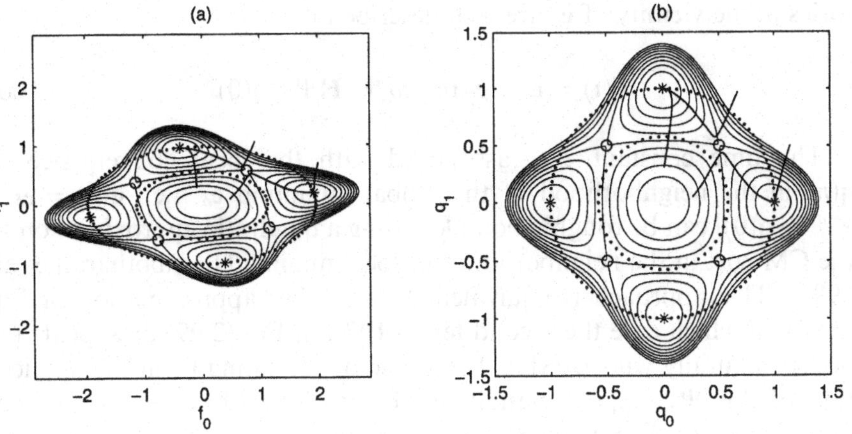

Figure 2.18 Exact gradient trajectories in (*a*) equalizer and (*b*) system space, with $L_f = 0$, $L_h = 1$ and BPSK. Both figures show dotted hyperannulus $\mathscr{S}$ boundaries and denote saddle points by $\circ$.

In the absence of noise, CMA's ODE can be derived from the mean CMA update term $E\{\mathbf{r}(n)\mathbf{f}^t\mathbf{r}(n)(\gamma - |\mathbf{f}^t\mathbf{r}(n)|^2)\} = \sigma_s^4\mathbf{H}\Delta_{\mathbf{q}}\mathbf{q}$ [recall Eq. (2.70)]. Transforming to the system space, the ODE becomes

$$\frac{d\mathbf{q}}{dt} = \mathbf{H}^t\frac{d\mathbf{f}}{dt} = \sigma_s^4\mathbf{H}^t\mathbf{H}\Delta_{\mathbf{q}}\mathbf{q} \tag{2.97}$$

The trajectories resulting from the solution of Eq. (2.97) represent average paths for the evolution of $\mathbf{q}(n)$ when updated by small step-size CMA. A display of some typical ODE trajectories is provided in Fig. 2.18.

Local Behavior Under the PBE conditions and in the vicinity of a CM minimum (i.e., $\mathbf{q} \approx \mathbf{e}_\delta$), the exact gradient can be approximated by the first term of its Taylor series expansion (Touzni and Fijalkow 1996) using

$$\Delta_{\mathbf{q}}\mathbf{q} = \underbrace{\Delta_{\mathbf{e}_\delta}\mathbf{e}_\delta}_{=0} - \Psi_{\mathbf{e}_\delta}(\mathbf{q} - \mathbf{e}_\delta) + o(\|\mathbf{q} - \mathbf{e}_\delta\|_2) \tag{2.98}$$

where it follows from Eq. (2.73) that, in the real-valued case,

$$\Psi_{\mathbf{e}_\delta} = (\kappa_g - \kappa_s)\mathbf{I} + 3(\kappa_s - 1)\mathbf{e}_\delta\mathbf{e}_\delta^T \tag{2.99}$$

Applying this approximation to Eq. (2.97), the exact gradient trajectories in the vicinity of $\mathbf{e}_\delta$ are well described by

$$\mathbf{q}(t) = \mathbf{e}_\delta + \exp(-t\sigma_s^4 \mathbf{H}^t \mathbf{H} \Psi_{\mathbf{e}_\delta})\mathbf{q}(0) \tag{2.100}$$

The time constant τ_{cma} associated with the local convergence of $\mathbf{q}(n)$ in the neighborhood of the global minimum $\mathbf{e}_\delta$ [i.e., $\mathbf{q}(n) = \mathbf{e}_\delta + e^{-n/\tau_{\text{cma}}}\mathbf{q}(0)$] can be roughly bounded using a quadratic approximation to the CM cost in the neighborhoods of local minima (Lambotharan et al. 1997). This approach is equivalent to a further approximation of Eq. (2.98), in which case the second term of $\Psi_{\mathbf{e}_\delta}$ in Eq. (2.99) disappears (as occurs naturally with BPSK). The trajectory resulting from the application of this $\Psi_{\mathbf{e}_\delta}$ approximation to Eq. (2.100) with $\sigma_s^2 = 1$ is characterized by a local-convergence time constant

$$\frac{1}{\mu(\kappa_g - \kappa_s)\lambda_{\max}(\mathbf{H}^T\mathbf{H})} \leq \tau_{\text{cma}} \leq \frac{1}{\mu(\kappa_g - \kappa_s)\lambda_{\min}(\mathbf{H}^T\mathbf{H})}$$

Here $\lambda_{\min}(\mathbf{H}^T\mathbf{H})$ and $\lambda_{\max}(\mathbf{H}^T\mathbf{H})$ indicate minimum and maximum eigenvalues of $\mathbf{H}^T\mathbf{H}$, respectively. Note that the term $(\kappa_g - \kappa_s)$ relates the steepness of the the locally approximated CMA cost surface to that of the LMS cost surface. For example, with a BPSK source, the local CMA surface curvature is twice that of LMS. Observe, then, that an increase in source kurtosis corresponds to a decrease in (local) convergence rate, where a Gaussian source provides the limiting case: zero convergence rate.

The locally quadratic approximation implies the following CMA step-size guideline (for $\sigma_s^2 = 1$), analogous to the LMS result in Eq. (2.60):

$$0 < \mu \leq \frac{2}{(\kappa_g - \kappa_s)\lambda_{\max}(\mathbf{H}^T\mathbf{H})} \tag{2.101}$$

Equation (2.101) gives a useful step-size guideline when considering trajectories in the vicinity of CM local minima, that is, near the point of convergence. In the quest for more general step-size guidelines, the quartic nature of J_{cm} [see Eq. (2.68)] suggests that a reasonable step-size upper bound should decrease for trajectories far from the origin. Specifically, Larimore and Treichler (1983) suggest that the upper bound should be equalizer dependent: inversely proportional to $\|\mathbf{f}\|_2^2$.

Global Behavior The multimodality of the CM cost function prevents a simple, direct comparison of time-to-convergence between CMA and LMS. An approximate understanding, however, may be obtained from the following two-stage convergence model for CMA:

1. Fast convergence to a region $\mathcal{S}$ near the unit sphere in system space (i.e., $\|\mathbf{q}\|_2 = 1$).
2. Within $\mathcal{S}$, slower convergence to a minimum.

Touzni and Fijalkow (1996) show that the norm of the mean gradient, $\|\mathbf{H}\mathbf{\Delta}_{\mathbf{q}}\mathbf{q}\|_2$, is very large when $\|\mathbf{q}\|_2 \gg 1$ (i.e., far from the unit sphere). Furthermore, in this region, the mean gradient points toward the origin. This implies that from an initialization $\mathbf{q}(0)$ far from the unit sphere, there is fast convergence toward the origin until $\mathbf{q}(n)$ penetrates the vicinity of the unit sphere. A similar effect occurs when $0 < \|\mathbf{q}(0)\|_2 \ll 1$. This time, the gradient points away from the origin so that there is fast convergence to the vicinity of the unit sphere from inside. These effects can be observed in Figs. 2.18 and 2.24.

A more precise definition of the aforementioned "region near the unit-sphere" is the hyperannulus $\mathcal{S}$ that includes all CM minima and saddle points. From Theorem 2.1, the CM power constraint (Zeng et al. 1998) implies

$$\mathcal{S} = \left\{ \mathbf{q} : \frac{\kappa_s}{3} \le \|\mathbf{q}\|_2^2 \le 1 \right\}$$

Figure 2.18 shows this region for the simplest case ($L_h + L_f + 1 = 2$ and $\kappa_s = 1$), where the borders of $\mathcal{S}$ are denoted by the dotted circles. For $\mathbf{q}$ in $\mathcal{S}$, the norm of the gradient is small and thus the convergence rate is slowed.

In Fig. 2.18, notice that as the trajectory $\mathbf{q}(t)$ approaches $\mathcal{S}$, it becomes attracted to the nearest stationary point. If the nearest attractor is a minimum, convergence heads toward the minimum at the rate determined by the approximately quadratic character of the region. If the nearest stationary point is a saddle point, the trajectory is first attracted to the saddle before converging to a stable minimum, consequently increasing convergence time (Lambotharan et al. 1997) as demonstrated by Fig. 2.24. (In higher dimensional cases, the trajectory can be attracted by several saddles before reaching a minimum—see Fig. 2.25.) Our brief observations about CMA's transient behavior highlight the importance of equalizer initialization, since it can be seen that convergence time is highly dependent on the trajectory's proximity to saddle points.

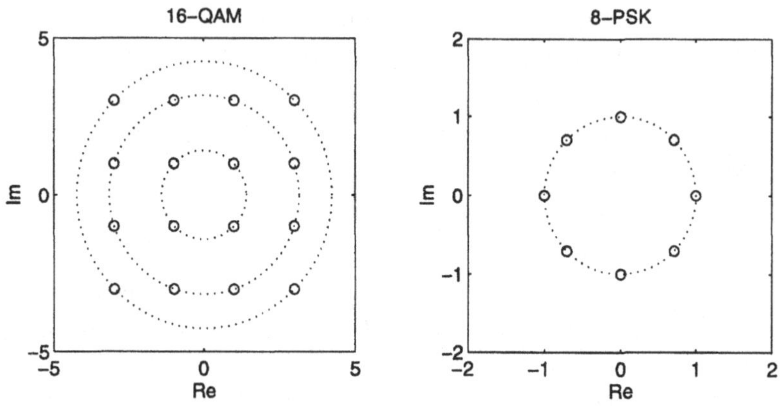

Figure 2.19 Nonconstant modulus source constellation (16-QAM) versus constant modulus source constellation (8-PSK).

Asymptotic Behavior: Performance at Steady-State We now analyze the steady-state scenario in which the CMA trajectory has converged to the vicinity of a CM minimum. In general, the CMA-adapted parameters continue to "jitter" around the true CM minimum due to a nonvanishing CMA update. This nonvanishing update may be a result of violations in the PBE conditions and/or the use of a non CM source (shown in Fig. 2.19).

We may consider quantifying the amount of parameter jitter in terms of *excess CM-cost*, defined as the steady-state CM cost above that achieved by the fixed local CM minimum. The excess CM-cost induced by CMA is intuitively similar to the excess MSE induced by LMS (see Section 2.2) in that both stem from the adapted parameters "bouncing around the bottom of the bowl," as illustrated by Fig. 2.20. An alternate and perhaps more useful method of quantifying CMA's parameter jitter is in terms of excess MSE. We define the EMSE of CMA as the difference between CMA's steady-state MSE and the MSE of the (fixed) local CM receiver.[15]

When the PBE conditions are satisfied, the CM receiver achieves perfect symbol recovery (i.e., $y_n = s_{n-\delta}$), and hence zero MSE. In this case, the excess MSE of CMA is equal to the total MSE of CMA and, thus, is relatively straightforward to calculate (see Fig. 2.20*a*). When the locations of the Wiener and CM receivers are distinctly different, how-

[15] Note that the extra MSE of CMA is different from the "extra MSE" of CMA used in Theorem 2.3.

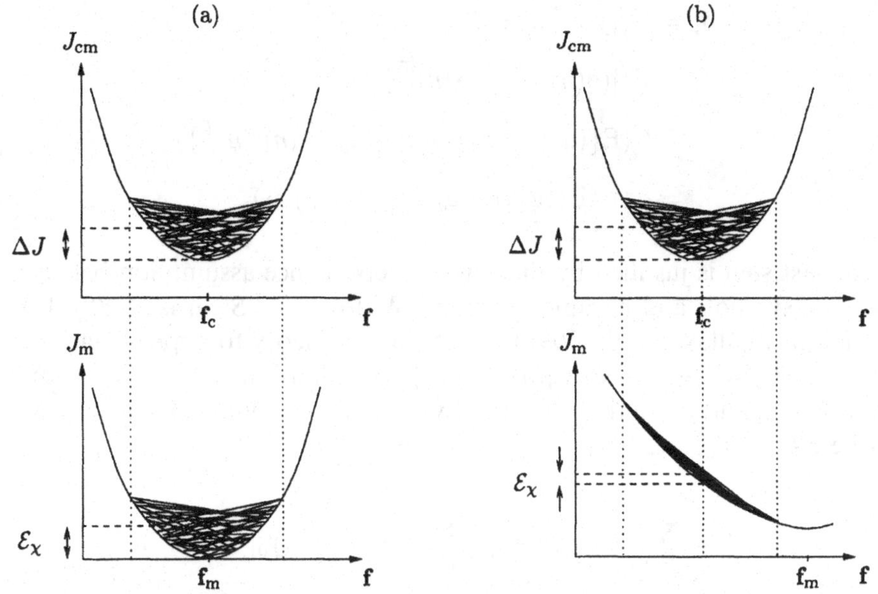

Figure 2.20 Illustration of the relationship between excess CM-cost ΔJ resulting from CMA adaptation and excess MSE $\mathcal{E}_\chi$ of the same CMA-adapted trajectory under (*a*) small and (*b*) larger violations of the PBE conditions.

ever, the steady-state EMSE of CMA is more difficult to calculate (see Fig. 2.20*b*). Fortunately, Godard's conjecture suggests that mild violations of the PBE conditions preserve the proximity between the CM and Wiener solutions, implying that EMSE derivations assuming PBE conditions are expected to hold suitably well in realistic (non-PBE) situations.

EMSE Due to a Non-CM Source As stated earlier, the CM receiver is capable of perfect symbol recovery when the PBE conditions are satisfied. Even so, the use of a non-CM source and a nonvanishing stepsize μ allows only mean-convergence of $\mathbf{q}(n)$ to $\mathbf{e}_\delta$. As evident from Eq. (2.96), a nonzero update will result from a fixed equalizer's inability to force the instantaneous squared modulus of a non-CM source to a constant (γ). Thus we expect the amount of parameter jitter to depend, in part, on particular source properties. Below, we approximate the steady-state excess MSE of CMA under satisfaction of the PBE conditions.

Following the approach adopted by Widrow and Stearns (1985), the EMSE at time index n can be calculated as

$$\begin{aligned}\mathscr{E}_\chi &= \mathrm{E}\{|y_n - s_{n-\delta}|^2\}\\ &= \mathrm{E}\{|(\mathbf{q}(n) - \mathbf{e}_\delta)^T\mathbf{s}(n)|^2\}\\ &= \mathrm{tr}(\mathrm{E}\{(\mathbf{q}(n) - \mathbf{e}_\delta)(\mathbf{q}(n) - \mathbf{e}_\delta)^T\mathbf{s}(n)\mathbf{s}(n)^T\})\\ &= \sigma_s^2\,\mathrm{tr}(\mathrm{E}\{(\mathbf{q}(n) - \mathbf{e}_\delta)(\mathbf{q}(n) - \mathbf{e}_\delta)^T\})\end{aligned}$$

The last step is justified by the usual independence assumption between source symbols and parameter errors (Widrow and Stearns (1985). For small μ, Fijalkow et al. (1998) use averaging theory to approximate the matrix $\mathbf{P} \triangleq \mathrm{E}\{(\mathbf{q}(n) - \mathbf{e}_\delta)(\mathbf{q}(n) - \mathbf{e}_\delta)^T\}$. At steady state, $\mathbf{P}$ is the unique positive-definite solution of the Lyapunov equation $\mathbf{QP} + \mathbf{PQ} = \mathbf{M}$, where $\mathbf{Q} = \Psi_{\mathbf{e}_\delta}$ and

$$\mathbf{M} = \sum_{n=-\infty}^{+\infty} \mathrm{E}\{y_n(\gamma - |y_n|^2)\mathbf{r}(n)(y_0(\gamma - |y_0|^2)\mathbf{r}(0))^T\}$$

[see Benveniste et al. (1990, 107)]. Under the PBE conditions and at steady state, the output y_n takes the form $y_n = \mathbf{e}_\delta^T\mathbf{s}(n)$, which, when used above, yields the following approximation for the EMSE of CMA (Fijalkow et al. 1998):

$$\mathscr{E}_\chi \approx \frac{\mu P(L_f + 1)}{2}\left(\frac{(\mathrm{E}\{s_n^6\}/\sigma_s^6) - {\kappa_s}^2}{\kappa_g - \kappa_s}\sigma_s^4\right)\sigma_r^2 \tag{2.102}$$

In Eq. (2.102), $\sigma_r^2 \triangleq \mathrm{E}\{|r_k|^2\}$ denotes the power of the received signal.

Equation (2.102), approximating the EMSE of CMA, looks similar to Eq. (2.63), which approximated the EMSE of LMS. Note the quantities present in each: step-size μ, total number of equalizer taps $P(L_f + 1)$, and received power σ_r^2. The EMSE component unique to CMA pertains to the source distribution, from which we make the following observations:

- EMSE is 0 for a CM source, that is, when $\kappa_s = 1$, when the PBE conditions are satisfied;
- EMSE increases as the source kurtosis approaches that of a Gaussian, that is, $\kappa_s \to \kappa_g$;
- EMSE is proportional to a sixth-order moment of the source, implying high sensitivity to non-CM source characteristics.

Our observations emphasize that nearly Gaussian source processes are particularly difficult for CMA-adapted blind equalization, which is commonly recognized. Specifically, as $(\kappa_g - \kappa_s) \rightarrow 0$, it can be shown that $((\mathrm{E}\{s_n^6\}/\sigma_s^6) - \kappa_s^2)$ remains nonzero [see, e.g., Rosenblatt (1985)] so that EMSE tends to infinity.

EMSE Due to Violations in PBE Conditions As stated earlier, a globally nonzero CM-cost implies a nonzero CMA update at all points in the equalizer space. This, in turn, results in steady-state parameter jitter around CM local minima and yields nonzero steady-state excess CM-cost. Recall that violation of any of the PBE conditions **(A1)–(A3)** yields the following:

- CM minima have CM cost greater than zero.
- Not all CM minima yield the same CM cost.
- CM receivers are no longer equivalent to MMSE or ZF receivers.

Hence, we conclude that any of the following—channel undermodeling, lack of subchannel disparity, the presence of channel noise, or the use of a temporally correlated source—will increase the steady-state level of excess CM-cost. An example is presented in Section 2.4, where the difference between the dashed and solid lines in Figs. 2.23*d–f* reveals EMSE due to noise, undermodeling, and nearly common subchannel roots.

The fact that the CM cost surface is nearly quadratic in the vicinity of local minima suggests that the excess CM-cost for CMA should be proportional to the same factors determining EMSE for LMS: step size, equalizer length, received signal power, and CM cost at the minimum. Unfortunately, translating this excess CM-cost into excess MSE is not always straightforward, especially for significant violations in the PBE conditions (recall Fig. 2.20). This is due in part to the fact that we lack closed-form expressions for the CM receivers when any of the conditions **(A1)–(A3)** is violated. To summarize, a method of quantifying the EMSE of CMA under arbitrary violation of the PBE conditions remains an open problem.

2.4 CMA-ADAPTED-EQUALIZER DESIGN ISSUES WITH ILLUSTRATIVE EXAMPLES

This section addresses design issues regarding fractionally spaced equalizers adapted by means of the CMA. In this context, equalizer design can be reduced to the selection of three key parameters: equalizer length, step size, and initialization. It will be shown that equalizer performance (e.g., MSE or convergence rate) is directly linked to each of these design parameters.

With this motivation, we seek to establish design guidelines for each parameter. When possible, we consider each parameter independently and follow an example-driven tutorial approach in developing these guidelines. The reader may notice that in many cases there are no clear answers to the optimal design choice. Despite the ambiguities, we are able to provide reasonable design guidelines supported by analysis and illustrative examples.

For the remainder of this section, we assume that the designer has a rough knowledge of channel characteristics in order to effectively choose the equalizer design parameters (i.e., blind equalization does not imply "blind" equalizer *design*). We focus on the use of $T/2$-spaced CMA as a blind start-up for eventual transfer to DD equalization. Finally, we assume that the signaling format is known by the receiver, as is usually the case in cooperative communication.

2.4.1 Initialization of CMA

Recall that the MSE cost, Eq. (2.21), is dependent on the choice of system delay. Hence, the objective of MMSE equalization can be considered the minimization of MSE over both the equalizer parameters *and* the system delay, δ. In systems that employ training, the system delay is directly linked to the designer's choice of training sequence delay (relative to a data-derived synchronization interval). In blind receivers employing CMA, *the system delay achieved by the steady-state equalizer is a direct result of the equalizer initialization.*

Take, for example, the simple 2-tap system of Fig. 2.21. This figure shows two CMA equalizer tap trajectories superimposed on the CMA cost contours that converge to different solutions for the channel $\mathbf{h} = (0.2, 0.5, 1.0, -0.1)^T$ at 10-dB SNR. Such convergence behavior can be understood by the equalizer-space regions of convergence, whose boundaries are plotted in Fig. 2.21 with dotted lines. The trajectory initialized within the region corresponding to CM minimum A achieved MSE $= 0.103$, while the trajectory initialized within the region corresponding to minimum B achieved MSE $= 0.294$. The CM minimum A is near the Wiener solution for delay $\delta = 1$, while B neighbors the Wiener solution for $\delta = 0$. Thus, for this channel/noise combination, we conclude that the system delay $\delta = 1$ is optimal. (The minima corresponding to the reflections of A and B through the origin correspond to sign in-

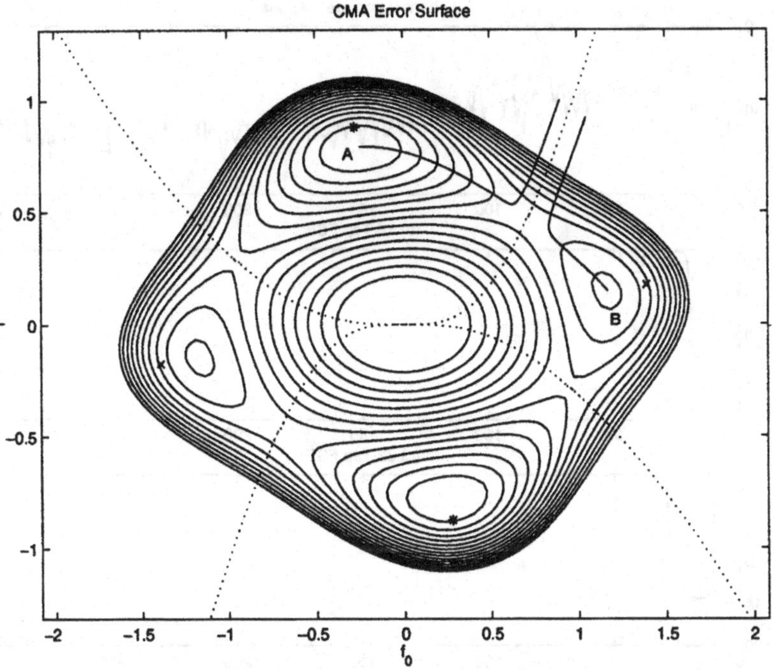

Figure 2.21 Two initializations of CMA with convergence to minima with different MSEs. Region-of-attraction boundaries denoted by dotted lines, and Wiener solutions for $\delta = 0$ and $\delta = 1$ represented by $(\times)$ and $(*)$, respectively.

versions of the system responses at A and B, respectively, as discussed in Section 2.3. Allowing for sign ambiguity, these reflected minima have equal performance.)

To understand the effects of initialization on equalizers with more than two taps, we need to consider the MMSE performance achieved at various system delays. Consider the delay-dependent performances in Fig. 2.22*b–e*, where minimum MSE is plotted versus system delay for FSE lengths of 16, 32, 64, and 299. The channel used for these experiments was SPIB[16] Microwave Channel 3, whose length-300 impulse-response magnitude appears in Fig. 2.22*a*. The MMSE traces demonstrate a behavior common to nearly all channel-equalizer combinations: system delays far from the channel cursor exhibit relatively poor performance. (The channel *cursor* is defined as the location of peak magnitude in the channel impulse response.) A reasonable conclusion is that, if

[16] The SPIB database resides at `http://spib.rice.edu/spib/select_comm.html`.

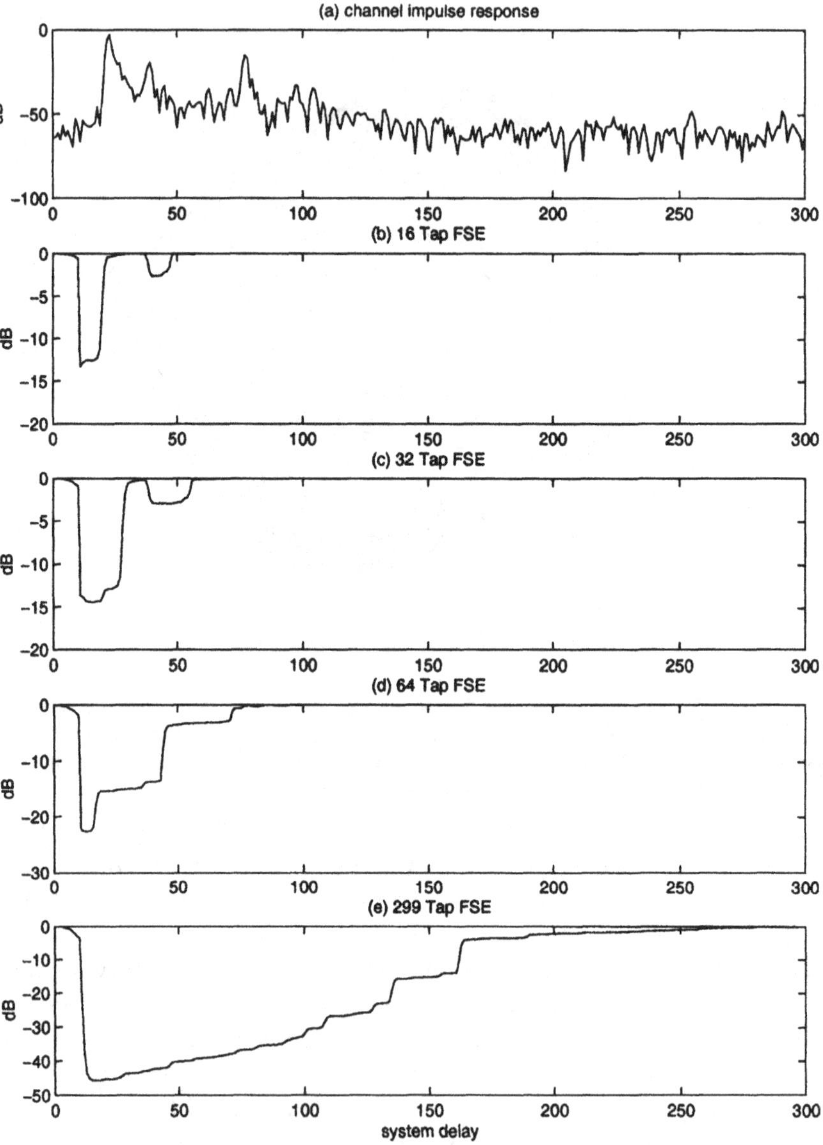

Figure 2.22 (*a*) SPIB Microwave Channel 3 impulse-response magnitude, and (*b*)–(*e*) achieved MMSE for various system delays and four equalizer lengths (SNR = 50 dB).

possible, *CMA should be initialized within the region of attraction associated with the MMSE-optimal system delay*. In practice, it is usually sufficient to initialize within the regions of attraction of minima having MSE below a particular application-dependent threshold.

When the equalizer length is sufficient to satisfy the PBE condition **(A1)**, the number of CMA minima (of a given phase) is equal to the number of achievable system delays, and so all delays are attainable through proper choice of initialization. As the length of the FSE decreases below the length required for perfect equalization, the number of CMA minima typically decreases as well, in which case fewer system delays remain reachable. This situation nearly always arises in practical environments, and the literature refers to such equalizers as being *undermodeled* with respect to the channel. As an example, Fig. 2.11 shows one CMA solution near two suboptimal Wiener solutions, while the other CMA solution remains near the Wiener receiver of optimal delay. Ideally, we would like the limited set of reachable delays to be those that achieve the best MMSE performance. Our claim is that undermodeled FSE-CMA typically provides this desired behavior, at least using popular initialization strategies. Before substantiating this claim, however, we will discuss these initialization strategies.

Single-Spike Initialization The most common strategy for initialization of CMA is termed *single-spike* initialization and derives its name from the baud-spaced strategy first suggested in Godard (1980). The single-spike technique specifies that all but one of the BSE coefficients are set to zero, though the location and amplitude of the nonzero tap has long been a subject of debate. Recently, Chung and Johnson (1998) made the observation that initializations with small norm result in almost certain CMA convergence to "good" local minima, that is, CM minima associated with relatively good MSE performance. Figure 2.21 illustrates this phenomenon for a 2-tap FSE: note how the regions of attraction corresponding to the better minima completely dominate the area near the origin. Unfortunately, the fact that the origin is a local maximum of the CM-cost surface implies that initializations with extremely small norms will result in slow (initial) convergence. It has been found that, in practical[17] situations, an initialization norm $\|\mathbf{f}_{\text{init}}\|_2 \approx 1$ is usually a good compromise between convergence rate and region-of-attraction selection.

The single-spike strategy is not always appropriate for fractionally spaced equalizers. For example, consider the trivial $T/2$-spaced channel $\mathbf{h} = (0, 1, 0, 0)^T$ with the single-spike FSE initialization $\mathbf{f} = (0, 1)^T$. If the equalizer output is decimated such that the odd-indexed samples are retained, then the equalizer output (and hence the CMA update) will be

[17] We assume that the received signal power is unit normalized by automatic gain control prior to equalization.

zero! In other words, the subchannel providing the signal energy will be nulled by the initial equalizer. What might be a more appropriate $T/2$-spaced initialization strategy is known as *double-spike* initialization and contains nonzero components in both subequalizers. Note that choosing the nonzero tap magnitudes $\approx 1/\sqrt{2}$ will result in an equalizer norm ≈ 1. It is customary to place the nonzero fractionally spaced taps adjacent to one another. This practice may be justified in the frequency domain, as adjacent positioning results in an initialization spectrum similar to a raised-cosine-pulse matched filter. The relative placement of the adjacent tap pair will be the focus of a later subsection.

Relationship Between FSE-Length and Initialization Sensitivity At this point we return to our discussion of initialization for undermodeled equalizers. (Recall that undermodeling nearly always exists in practice.) Previously we claimed that, under a reasonable initialization, FSE-CMA achieves a "good" range of reachable system delays. Using the unit-norm double-spike initialization strategy, we now offer example evidence for this claim.[18] Figure 2.23 shows steady-state MSE performance achieved by CMA for the full range of (adjacent) double-spike locations. For ease of comparison, the same channel and equalizer lengths used in Fig. 2.22 are used. Other details include the use of quarternary phase-shift keying (QPSK), an SNR of 50 dB, step sizes $\{0.005, 0.005, 0.0025\}$ for respective FSE lengths $\{16, 32, 64\}$, and simulation lengths of 10^6 iterations. Figure 2.23 has two important messages:

- First, the only system delays reached by FSE-CMA under double-spike initialization were indeed those corresponding to the low-MMSE "troughs" in Fig. 2.22; FSE-CMA avoided convergence to system delays characterized by poor MMSE. (Notice that the dashed lines in Fig. 2.23*d–f* correspond to the lowest sections of the MMSE traces in Fig. 2.22.)
- Second, the relative MSE performance versus delay seems to "even out" as the degree of undermodeling increases. In other words, initialization sensitivity *decreases* as equalizer length decreases. Note, however, that for the same (nonzero) noise level, the MMSE performance achieved by shorter equalizers is typically[19] worse. In

[18] As no rigorous proofs of optimality exist for the single/double-spike initialization scheme, we are not claiming that the reachable set of delays is *always* optimal. However, practitioners have found these initialization strategies to work well.

[19] Due to the effects of EMSE, longer adaptive equalizers will not always result in lower steady-state EMSE, as discussed later in this section and illustrated by Fig. 2.28.

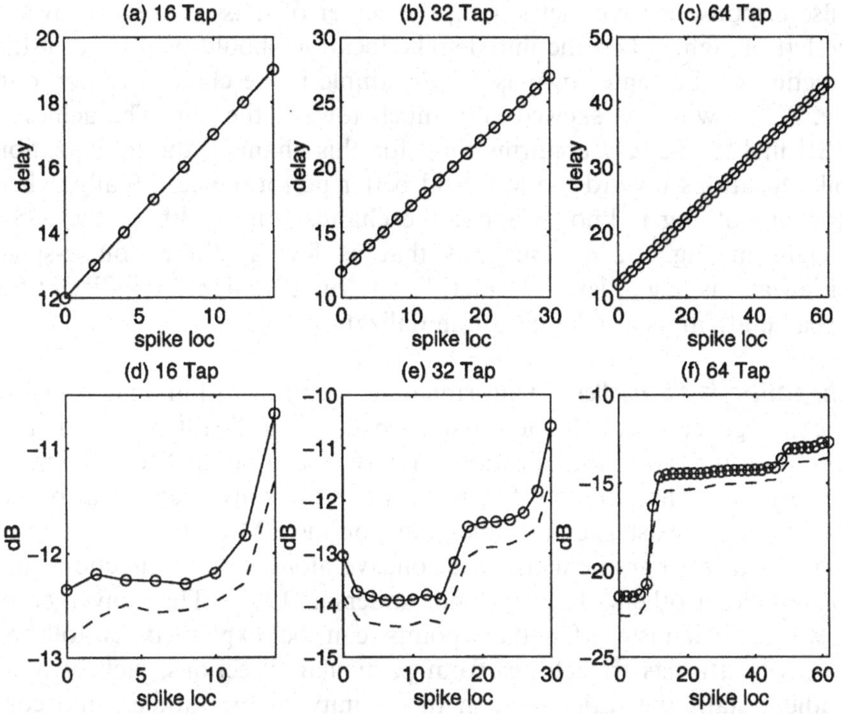

Figure 2.23 (*a*)–(*c*) Achieved system delay as a function of double-spike location, and (*d*)–(*f*) CMA's corresponding MSE performance for the channel response in Fig. 2.22*a*. The dashed lines indicate Wiener (i.e., MMSE) performance at the respective system delays.

fact, the best MSE achieved by the 16-tap CMA-adapted equalizer in Fig. 2.23 is not as good as the worst MSE achieved by the 64-tap equalizer.

The shape of the typical "trough" in the MMSE-versus-delay profile (Larimore et al. 1997) implies that, even for long equalizers, good performance is assured when the initialization spike is chosen anywhere within the region corresponding to the lowest-MMSE delay.

Spike-Delay Selection We note from Fig. 2.23*a–c* that there seems to be a direct correspondence between initial spike location and achieved system delay. A similar observation was made in Li and Ding (1995) for single-spike initialized BSEs. Thus, knowing something about the channel impulse-response shape can help with the selection of initialization. For example, if the channel impulse response is reasonably symmetric, a center spike is most appropriate. If, on the other hand, the channel impulse response is asymmetric (i.e., the center of mass is skewed toward the left or right), then the initial spike location should be moved in the direction of the center of mass. An example is the channel response in Fig. 2.22*a*, which is skewed very much toward the left. The achieved MSE in Fig. 2.23*d–f* confirms that, for this channel, the initialization spike locations toward the left yield better performance. Finally, when little or nothing is known about the channel, the width of the MSE troughs in Fig. 2.23*d–f* suggests that as few as three double-spike initializations (e.g., "left," "center," and "right") might be sufficient for repeated attempts at a successful initialization.

Attraction by Saddles Unfortunately, location within a good region of convergence is not the only requirement of an ideal CMA initialization. As it turns out, initialization can have a profound effect on time to convergence. The potential for arbitrarily slow convergence can be explained by the existence of saddle-points on the CM-cost surface, that is, points with zero gradient that are concave along certain directions and convex along others (Johnson and Anderson 1995). The convergence-slowing mechanism of saddle points can be explained as follows: concavity attracts trajectories from particular directions, lack of local gradient stalls the trajectories in the vicinity of the saddles, and convexity eventually ejects the trajectories in other directions.[20] Figure 2.24 shows examples of this behavior for different 2-tap FSE initializations using the noiseless $T/2$-spaced channel response $(0.2, 0.5, 1, -0.1)^T$ and a BPSK source. Note that in this two-dimensional example, the CMA trajectories are only capable of passing through one saddle before converging to a stable minimum. Figure 2.25 plots the CM-cost evolution for a higher-dimensional case in which the CMA trajectory passes near multiple saddles (Lambotharan et al. 1997). Here the $T/2$-spaced channel was $\mathbf{h} = (0.7571, -0.2175, 0.1010, 0.4185, 0.4038, 0.1762)^T$ and the 4-tap FSE was double-spike initialized at $\mathbf{f} = (1, 1, 0, 0)^T/\sqrt{2}$. Notice that even the popular double-spike initialization is susceptible to attraction by saddles.

Indirect Initialization Techniques If needed, the information gained from convergence to one CMA solution can be used to better initialize a second attempt. For instance, Tong and Zeng (1997) proposed an iterative scheme whereby a delay-shifted version of a channel estimate

[20] Technically speaking, the directions of attraction correspond to the eigenvectors of the Hessian associated with positive eigenvalues, while the directions of repulsion correspond to the eigenvectors associated with negative eigenvalues.

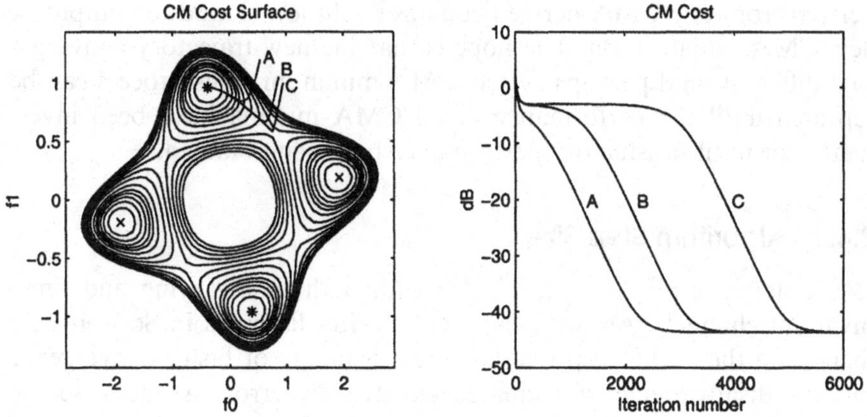

Figure 2.24 Two-tap FSE-CMA trajectories and CM-cost history for several initializations, showing the effect of a saddle point on convergence speed.

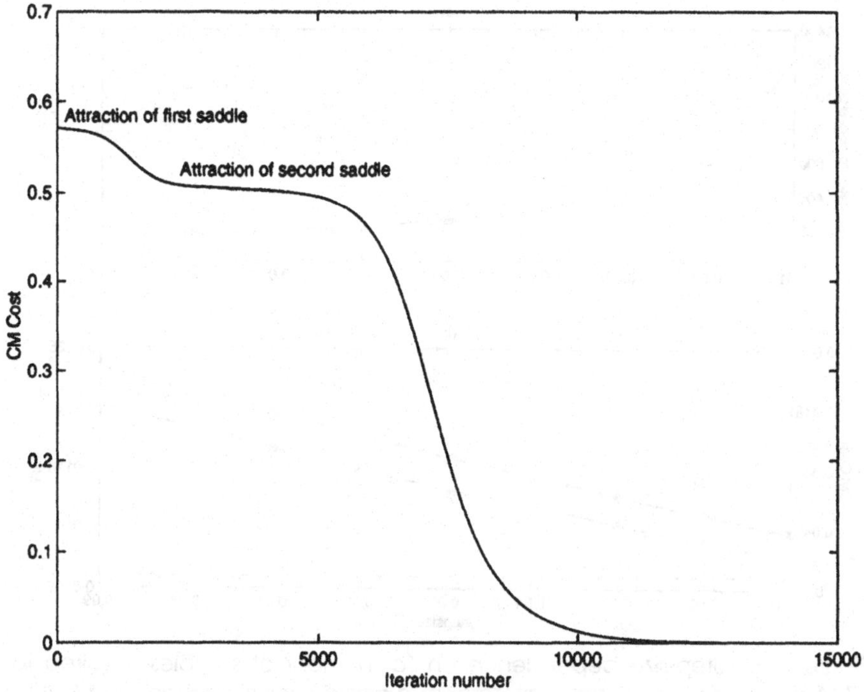

Figure 2.25 Slow convergence of a double-spike-initialized 4-tap FSE due to multiple saddle-point attractions.

derived from one CMA-derived equalizer solution is used to compute a new CMA initialization. The hope is that the new trajectory converges to a different (and perhaps better) CMA minimum. This process can be repeated until the performance of all CMA minima have been investigated or until satisfactory performance has been obtained.

2.4.2 Algorithm Step Size

CMA step size plays a critical role with both time-varying and time-invariant channels. As is the case with LMS (discussed in Section 2.2), increasing the CMA step size has the advantage of faster convergence but the disadvantage of increased steady-state error. As discussed in Section 2.3, this asymptotic performance may be quantified in terms of either excess CM cost or excess MSE. Figure 2.26 depicts, for a range of step sizes, the convergence time and the averaged steady-state values of CM cost and MSE. The experiments were conducted using BPSK, the channel $\mathbf{h} = (0.2, 0.5, 1.0, -0.1)^T$ at 20-dB SNR, and a 2-tap FSE

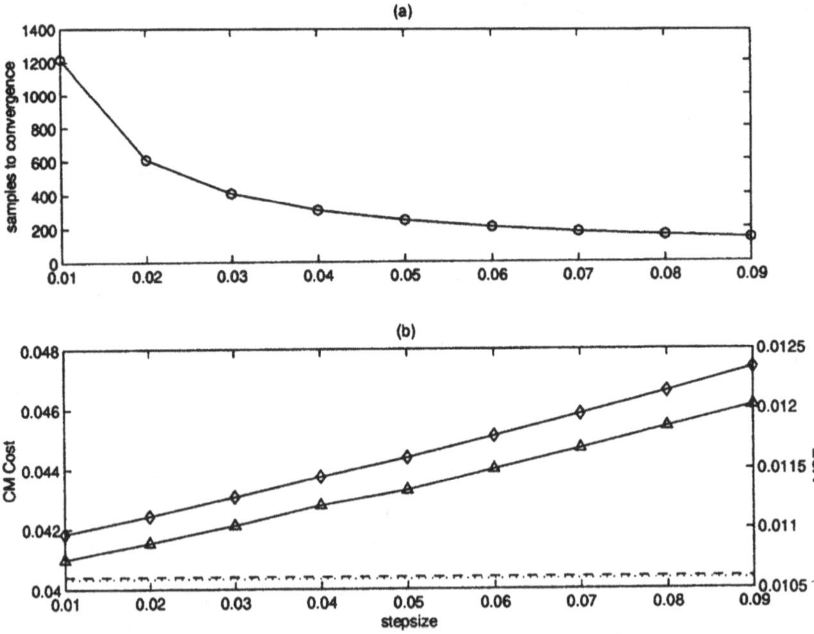

Figure 2.26 Step-size dependence on (*a*) number of samples required for CMA convergence to steady-state from a particular initialization, and (*b*) the resulting steady-state CM cost (◇) and MSE (Δ). Dashed line indicates (locally) minimum CM cost, while dotted line indicates (locally) minimum MSE.

initialized at $(1, 1)^T$. Evident is the fundamental trade-off between convergence rate and excess steady-state error for a time-invariant channel, as discussed in Section 2.3.

When the channel response is time varying, we would like the equalizer to adapt quickly enough to track the channel variations. The amount by which the adapting equalizer lags behind the optimal solution leads to a proportional increase in excess CM-cost (and, we expect, in excess MSE). Because a larger step size leads to faster adaptation, we expect that it decreases the lag contribution to CM cost. However, we have also seen that a larger step size increases the excess CM cost resulting from a non vanishing parameter update term (i.e., "adaptation noise"). Hence, we anticipate a compromise between the effects of tracking lag and gradient noise.

In practice, the guidelines for CMA step size may be obtained by slight modification of LMS step-size guidelines. First, CMA convergence rate in the vicinity of a "good" local minimum is known to be well-characterized by quadratic convergence models and be $\kappa_g - \kappa_s$ times as fast as LMS convergence rate (recall the material presented in Section 2.3). Note that, for uniformly distributed PAM and QAM constellations, we may show $0.6 < \kappa_g - \kappa_s \leq 2$. Second, the EMSE of CMA and LMS differ [recall Eqs. (2.63) and (2.102)]. Though both are linearly proportional to step size, equalizer length, and received signal power, a non-CM source will increase CMA's EMSE significantly above that of LMS (all else being equal). However, recalling that the goal of CMA is typically the achievement of MSE below the DD transfer level (as seen elsewhere in this section), particularly low values of steady-state EMSE might not be necessary. We conclude that CMA's step size may often be chosen within a factor of 2 of the recommended LMS step size (e.g., the values in Treichler et al. (1996), so long as the anticipated value of the (source-dependent) EMSE is deemed adequate.

In order to demonstrate the step-size trade-off for FSE-CMA, we use a random-walk channel with impulse response $\mathbf{h}(k+1) = \mathbf{h}(k) + \mathbf{n}(k)$ at time $(kT/2)$. The vector process $\{\mathbf{n}(k)\}$ is composed of zero-mean uncorrelated Gaussian elements with variance 0.0005. With an additive channel noise of $\sigma_w^2 = 0.05$, averaged steady-state CM-cost is plotted in Fig. 2.27 for various values of step size.

2.4.3 Equalizer Length

Equalizer length is one of the most important variables in communication receiver system design. It is directly related to both optimum receiver performance and implementation cost. For example, approxi-

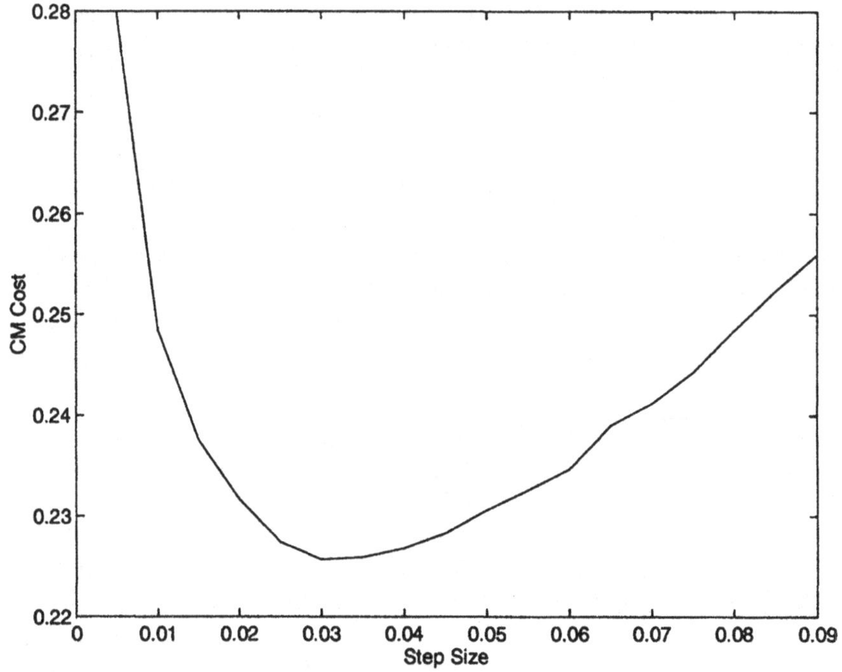

Figure 2.27 Averaged steady-state CM cost versus step size for random-walk channel $\mathbf{h}(k)$.

mately 80% of Applied Signal Technology's very-large-scale integrated (VLSI) chip described in Treichler et al. (1996) is devoted to the FSE.

A common myth regarding equalizer length selection is that a longer FSE always achieves better MSE performance than a shorter FSE. This principle is false for equalizers adapted by a SGD algorithm such as CMA. In the presence of channel noise, a longer FSE is better at mitigating the effects of ISI and noise than a shorter FSE, but a longer FSE will incur the penalty of additional MSE due to increased adaptation noise (Widrow et al. 1976). Although CMA and LMS both exhibit this undesirable excess MSE (EMSE) due to adaptation noise (see Section 2.2), CMA suffers additional EMSE when used with a non-CM source. (Recall from Section 2.3 that a non-CM source makes it impossible to achieve zero CM cost, and it is this nonzero CM cost that causes the CMA-adapted FSE coefficients to "rattle around" the locally optimal CM solution when driven by a nonvanishing step size.) The resulting EMSE under the PBE conditions was approximated in Eq. (2.102) and, as with LMS in Eq. (2.63), is proportional to both step size and FSE length. Herein lies the classic compromise: the FSE length should be

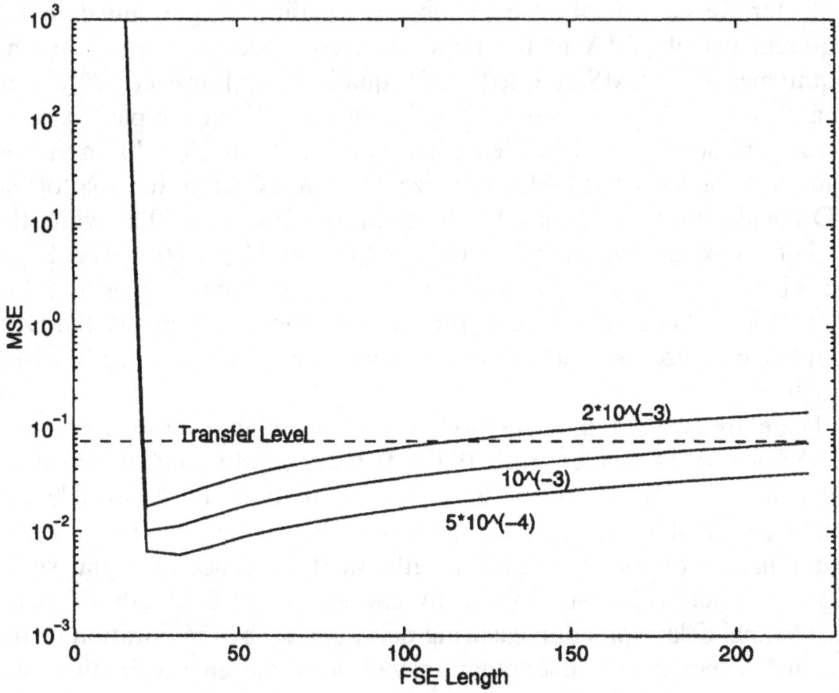

Figure 2.28 MSE performance of CMA as a function of FSE length for various step sizes: SPIB Microwave Channel 2, 16-QAM, no noise.

long enough to mitigate channel-induced noise and ISI, but short enough to prevent the MSE due to adaptation noise from dominating.

Combining the approximate effects of equalizer length and source kurtosis on MSE from Endres et al. (1997b) and Eq. (2.102), respectively, Fig. 2.28 presents steady-state MSE performance of CMA for various equalizer lengths and step sizes. The simulations assume a 16-QAM source over SPIB Microwave Channel 2 in the absence of noise. Our goal for the blind start-up mode is to choose a length/step-size combination capable of achieving an MSE below the DD transfer level. The dashed line in Fig. 2.28 indicates this transfer level for 16-QAM. (Choosing the transfer level will be the subject of the next section.) Figure 2.28 demonstrates that a longer equalizer does not necessarily provide better steady-state performance than a shorter one. In fact, with certain channel/step-size combinations, an FSE-length *less* than that required for perfect equalization may be required for CMA to reach the DD transfer level. A similar observation was made for BSE-CMA in Li and Liu (1996).

So far we have focused on choosing equalizer length suited to the requirements of CMA in the blind start-up mode. The behavior and requirements of LMS-adapted DD equalization, however, may lead to a slightly different equalizer-length trade-off. For example, we have already noticed that LMS equalizers are not affected by non-CM sources in the way that CMA equalizers are. In addition, the goal of the DD equalization mode may be to attain an SER of $\leq 10^{-6}$, while the goal of CMA equalization might only be to attain SER $< 10^{-2}$. The differences between the start-up and DD modes suggest that it may be useful to allow different equalizer lengths for the two modes. In general, the DD-adapted equalizer should be longer than the CMA-adapted start-up equalizer.

There are two common methods for determining a reasonable guess at FSE length (Treichler et al. 1998). If representative signal snapshots are available, one approach is to run simulations of various length equalizers on these signal snapshots in both CMA and DD modes to determine an optimal equalizer length. In the absence of signal snapshots, a good rule-of-thumb is to choose an FSE length of twice the channel delay spread, measuring delay spread by the duration of the channel response region containing 99.9% of the energy in the total channel response (Treichler et al. 1996). This rule of thumb generally applies to communication systems with moderate complexity constellations (e.g., 64-QAM). Higher complexity constellations may require a longer equalizer to achieve reasonable performance.

2.4.4 DD Transfer Level

Used for blind start-up, the role of CMA becomes one of reducing the SER to a level sufficient for successful DD equalization. Practitioners usually consider SER between 10^{-1} and 10^{-2} as reasonable for initiating DD equalization. Since we often measure CMA performance in terms of MSE, we would like a means of translating the SER requirement to an MSE requirement.

In general, the exact relationship between SER and MSE is not simple to describe or compute, since it depends on the probability density function (pdf) of the interference (Gitlin et al. 1992). One approximation considers the ISI as an additional source of additive white Gaussian noise. Though not a good assumption in general, it is adequate for obtaining design guidelines based on DD transfer level. Recall the expression for the steady-state MSE of LMS given in Eq. (2.62). If the signal-to-interference (SIR) ratio is given as $\Upsilon = \sigma_s^2/\mathrm{E}\{|e_n|^2\}$, then under the approximation of Gaussian interference the SER may be expressed as

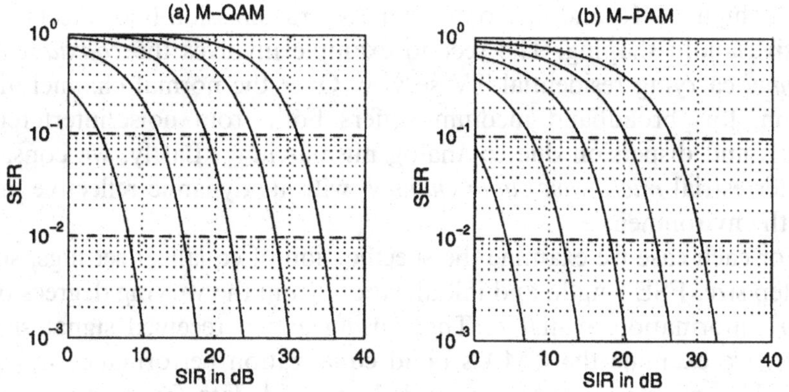

Figure 2.29 Approximate SER versus SIR for (*a*) M-QAM alphabets, $M = \{4, 16, 64, 256, 1024\}$ from left to right, and (*b*) M-PAM alphabets, $M = \{2, 4, 8, 16, 32\}$ from left to right.

$$\text{SER} = 1 - \left[1 - \left(1 - \frac{1}{\sqrt{M}}\right) \text{erfc}\left(\sqrt{\frac{3\Upsilon}{2(M-1)}}\right)\right]^2 \tag{2.103}$$

for an M-QAM system and as

$$\text{SER} = \frac{(M-1)}{M} \text{erfc}\left(\sqrt{\frac{3\Upsilon}{M^2 - 1}}\right) \tag{2.104}$$

for an M-PAM system. Figure 2.29 illustrates the SER versus SIR approximation for various source alphabets.

2.4.5 Conclusions

It should be clear from the examples and analysis of this section that the design choices for initialization, step size, and equalizer length all impact receiver performance. Most of these design choices rely on knowledge of the prototypical channel impulse response, or at least a rough notion of the channel class.

2.5 CASE STUDIES

To illustrate design methodology, we consider three actual communication scenarios addressing different classes of channel impairments to digital transmissions. The first, the *voiceband telephone channel*, typically has a high SNR and is dominated by transmission-line effects and passive audio shaping. The second example involves the *coaxial-cable channel* carrying residential TV service. Over the normal channel allocation, this broadband medium suffers both from noise/interference effects and dispersion due to analog mismatching. Lastly, we consider the terrestrial *microwave radio channel*, with its dynamic reflective multipath environment.

For each case we examine the specification of equalizer settings, such as step size, FSE length, and initialization, from the varying degrees of *a priori* information available. Then, using actual received signal snapshots, we examine the CMA's blind equalization performance in each case. Use of experimentally acquired received data from actual communications channels also requires us to consider other issues such as carrier-frequency offset, noise, and channel time-variations. The effects of carrier-frequency offset, to which the CMA is immune because of its independence of phase, will be illustrated in the voiceband modem example. The digital-cable television example will illustrate the effects of noise on blind equalization of large signal constellations. The last example, mobile microwave communication, is characterized by a time-varying channel. We will show how the three signal environments impact our equalizer parameter selection and equalizer performance.

Good design practice requires us to define the equalizer using any information known about the particular signal environment. For example, for a particular communications link, we may be able to characterize a suitable range of equalizer lengths, communication channel delays, degree of time variation in the channel, and range of SNR. To minimize hardware complexity and maximize agility, we would like to simultaneously choose the shortest equalizer length, largest step size, and most broadly successful initialization. While our objective is to rapidly reduce the ISI to allow carrier tracking and switching to DD equalization in the least amount of time, practically we try to select the parameters to provide a solution allowing steady-state DD equalization within the constraints of the software and/or hardware being used to process the data.

In the following examples, all of the channels are derived from empirical measurements by Applied Signal Technology, Inc., and reside in the SPIB database at `http://spib.rice.edu/spib/select_comm.html`.

2.5.1 Voiceband Modem: The Dial-Up Telephone Channel

First we consider a class of channels that exhibits a wide range of challenging characteristics. Today, data transmission over a dial-up telephone channel is taken for granted as an integral part of personal telecommunications, enabling, for example, the individual's access to the Internet as well as widespread use of dial-up facsimile machines. The channel effects in such a transmission medium are due mainly to the analog "local subscriber loop" originally engineered for speech service. This passive connection runs from the subscriber's premises to the telephone service provider's local office facility. The signal must be carried over this copper line, up to 5 km in length, subject not only to passive transmission-line dispersion, but also to a variety of nonideal effects, such as spliced wire gauges, nonterminated shunts, and dielectric variations. Once connected to the public switched telephone network (PSTN), the signal occupies the voice band in the range of 300 to 3400 Hz, with band-limiting sufficient to support the 8-kHz sampling operation used for TDM in the digital switching and transmission system. The channel typically exhibits phase irregularities near band edges, with a sloping or rippling magnitude loss on the upper half of the passband; Fig. 2.30*a* shows the magnitude response for a typical voiceband channel.

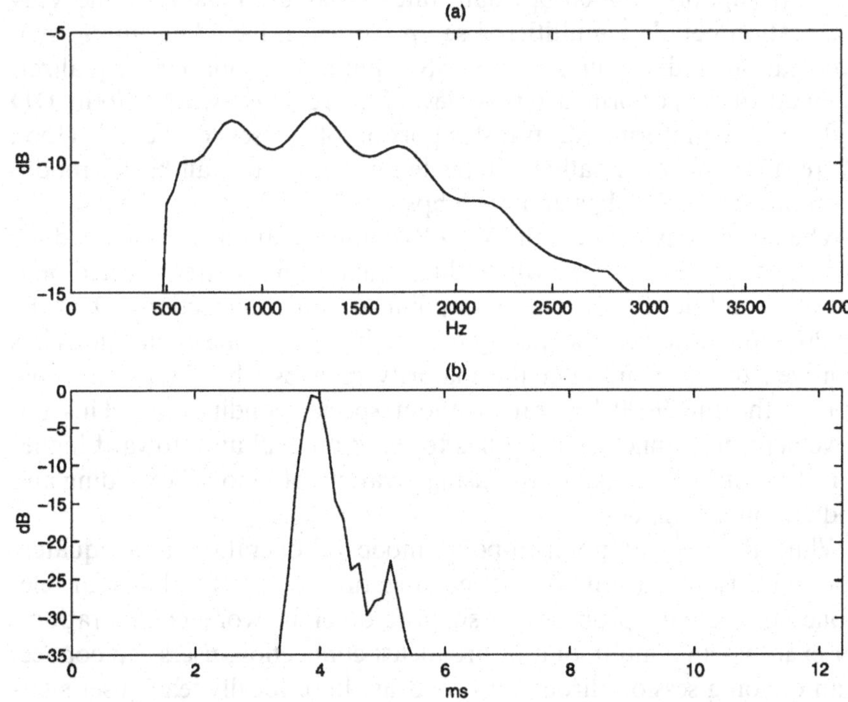

Figure 2.30 Prototypical voiceband channel: (*a*) frequency response magnitude; (*b*) impulse response magnitude.

Assuring reliable transmission of data under such conditions has proven very challenging and has motivated some of the earliest endeavors in adaptive equalization and coded modulation. The nature of the basic channel limits symbol rate to on the order of 3400 symbols/s, with modulating carrier frequency between 1700 and 1900 Hz. Furthermore, the complexity of QAM modulation, hence the actual data rate, is limited mainly by severe ISI rather than additive noise, since the SNR typically exceeds 30 dB. Because the channel characteristics may vary widely from connection to connection, equalization of so-called "dial-up" service by the receiving terminal must be highly robust. This often involves automatic training options that not only initialize equalizer parameters but also allow adjustment of modulation symbol rate and constellation complexity when conditions dictate.

Currently available voice-channel modems rely on training at the start of a connection. For many years, the premier modulation technique for voice-channel service was V.29, formalized in a published recommendation (ITU 1988) by the International Telecommunication Union (ITU). It uses a differentially encoded QAM constellation transmitted at 2400 symbols/s. At this highest rate, the 16-point constellation depicted in Fig. 2.31 supports a 9600 bps data rate. To assure reliability, the V.29 specification details a multitiered setup sequence, including timing synchronization and training of an MSE-minimizing adaptive equalizer. When adequate performance is achieved, the receiver switches to its DD mode, and the information transfer portion of the session begins. Upon failure, the procedure calls for repeated attempts at "fall-back" modulation modes of 7200 bps and 4800 bps.

When originally defined, the V.29 Recommendation was intended for 4-wire or leased service, where high-quality lines are selected and specially conditioned for long-term point-to-point connections. Over the last 20 years, however, the quality of switched telephone connections has improved to the point that the majority of subscriber loops can now support the full 9600 bps rate without special conditioning. This improvement in channel capacity has fostered an evolution toward higher point-to-point service data rates using a variety of innovative coding and modulation techniques.

While the present point-to-point mode of operation is adequately supported using trained MMSE equalization, future flexibility in telephone connectivity promises to support other network configurations. For example, in a multipoint or broadcast connection, users can connect to an ongoing session through a broadcast hub. Ideally, each user's terminal will establish receiver lock autonomously so as not to unduly impact the overall data rate (as would occur with the inclusion of training).

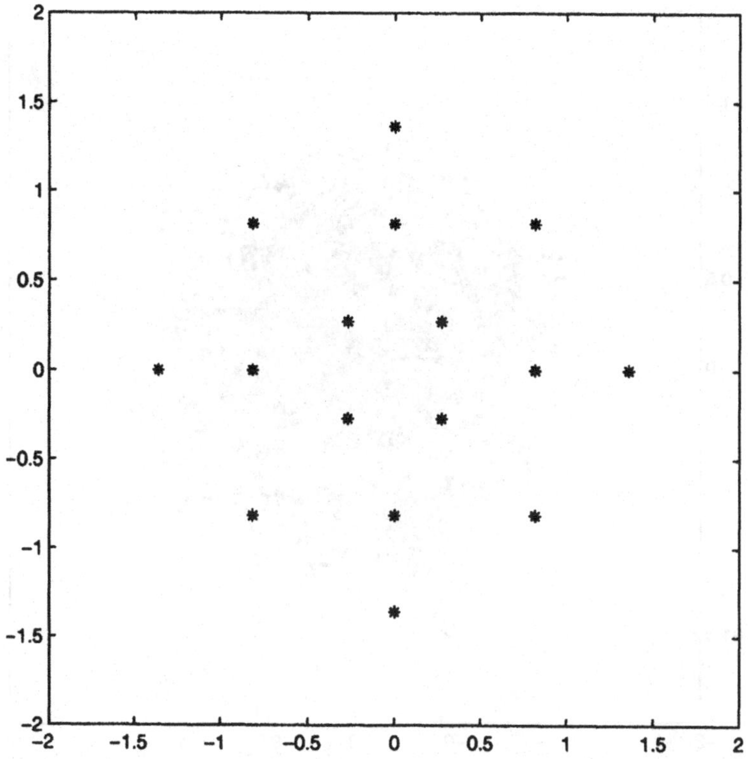

Figure 2.31 V.29 constellation.

Below we consider the design implications for a blind equalizer suitable for this type of broadcast connection over existing telephone transmission facilities. The goal for such a blind equalizer is to reduce constellation dispersion to the point where a DD scheme can reliably converge to the MMSE equalizer solution. Note that, while we illustrate using the V.29 modulation format, this by no means represents a limit on the data rate.

The typical voice channel has the frequency response magnitude shown in Fig. 2.30*a*, and has the impulse-response magnitude seen in Fig. 2.30*b*, where the bulk of the energy is confined to a time duration on the order of 2–3 ms. For our design, we recall the equalizer length discussion in Section 2.4 and select an equalizer duration of 30 $T/2$-spaced taps (spanning 6.25 ms, or about twice the channel spread). For adequate agility, we specify a step size $\mu = 0.001$, providing a basic response time in the millisecond range.

The performance of this prototype design can be demonstrated using a record of actual V.29 signals captured at a subscriber's premises

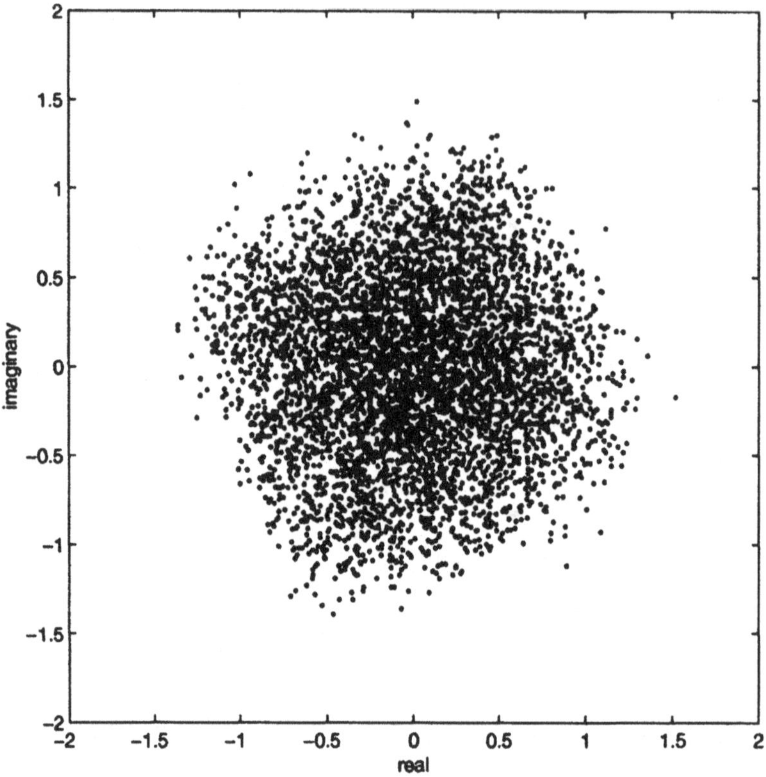

Figure 2.32 *T*/2-spaced input in V.29 demodulation example.

(as found in the SPIB database). The waveform is quadrature down-converted by the known carrier frequency of 1700 Hz and sampled at a $T/2$-spaced rate of 4800 samples/s. The resulting complex baseband samples are plotted in Fig. 2.32, where the indistinct cloud indicates complete closure of the eye of the data waveform. Following convergence of the blind equalizer, the samples take the form shown in Fig. 2.33, clearly exhibiting clustering consistent with the V.29 constellation. At this point, a phase-tracking loop is used to estimate bulk phase error and properly align the constellation with the decision regions. Figure 2.34 shows the end result following switchover to the DD mode.

It is interesting to examine Fig. 2.33, the constellation following blind equalization, more closely. The elongated shape of the individual cluster points is symptomatic of "phase roll" due to error in the estimation of carrier frequency. In this case, the accumulated phase change amounts to only about 20° over the observation window. In many cases the carrier offset may be several Hz [e.g., in analog frequency-division multi-

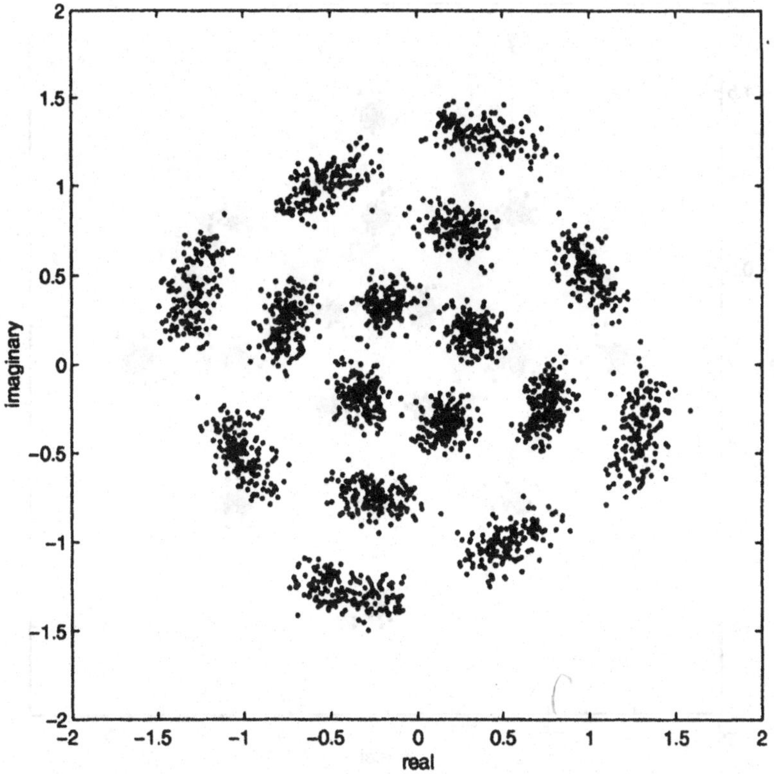

Figure 2.33 Output of CMA-adapted equalizer in V.29 demodulation example.

plexing (FDM) transmission], resulting in a spinning of the equalizer output constellation, as illustrated in Fig. 2.35. Because its adaptation depends only on modulus error, CMA-based blind equalization is largely immune to such phase drift and frequency error. By decoupling mitigation of channel effects from carrier phase-tracking circuitry in this manner, accurate carrier phase recovery is greatly simplified.

The specifics of this modem design are based on dispersion exhibited by a typical voice channel and serve as a feasibility baseline for the extension of traditional voice-channel technology. The benefits possible through blind recovery enable modem usage in the broadcast environment of the future.

2.5.2 Digital TV: The Broadband Coaxial-Cable Channel

Next, we focus on the use of coaxial cable as a transmission medium for broadcast distribution of digital television signals for residential cover-

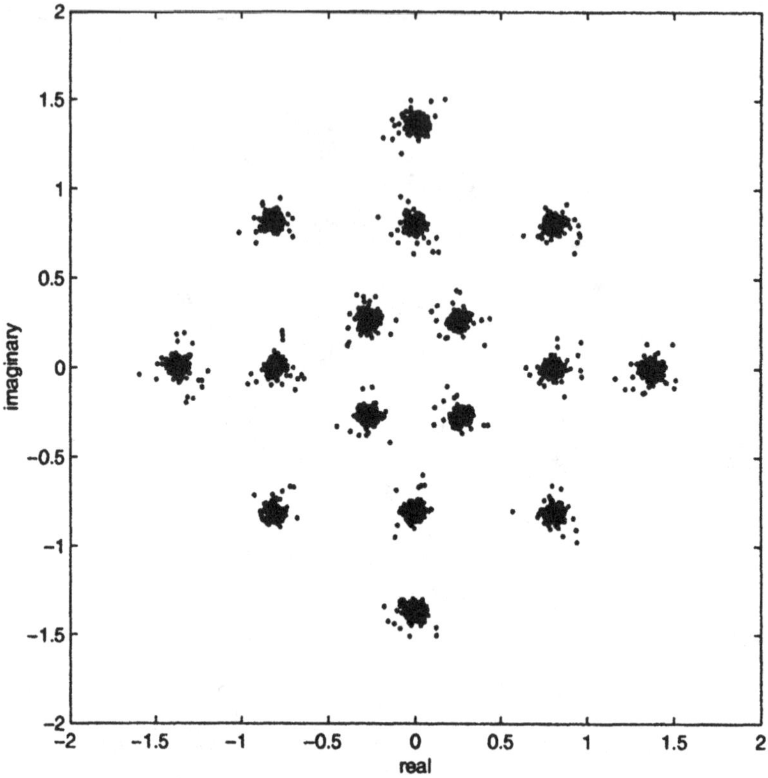

Figure 2.34 Output after DD-equalization and carrier tracking in V.29 demodulation example.

age. Cable service providers operate regional "head end" distribution centers, where national and international feeds are combined with local programming into a composite signal and conditioned for transmission to local subscribers. Much of the physical cable currently in place was originally intended for analog transmission, that is, 6-MHz TV signals frequency-multiplexed to channel assignments of up to 700 MHz. Relative to broadcast transmission, distribution via cable provides a potentially higher-quality signal, unaffected by atmospheric attenuation, RF interference and multipath, and where frequency-dependent losses are normally small throughout any given 6-MHz band. Practical cable systems, however, do suffer from nonideal effects related to reflective mismatched terminations, imperfect isolation, and crosstalk. In addition, shielding breaks or weak connections may allow so-called "ingress" interference to enter the band (from atmospheric EMI and/or RF broadcast signals).

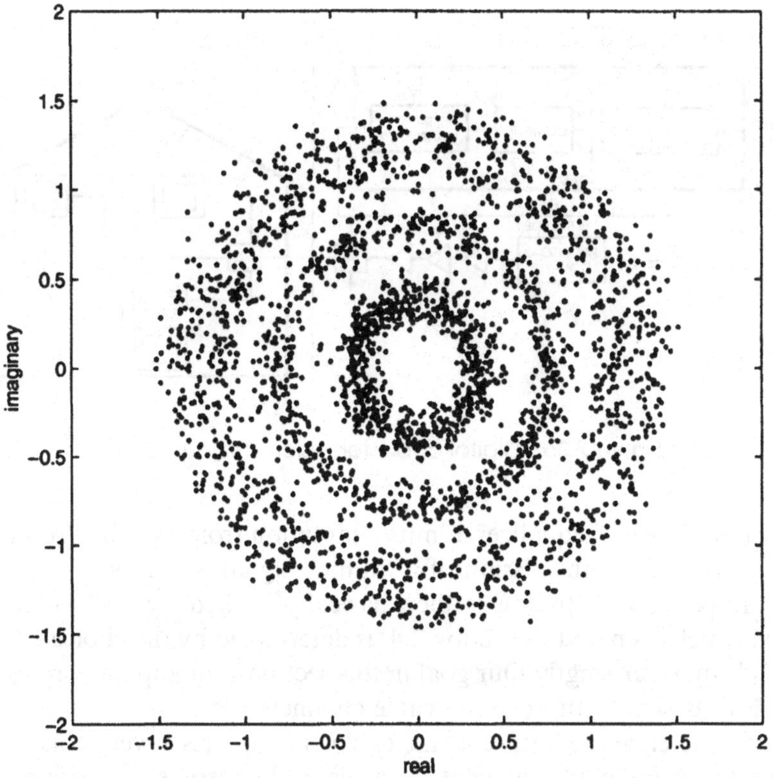

Figure 2.35 Output of CMA-adapted equalizer under 2 Hz of carrier frequency offset in V.29 demodulation example.

To provide better signal quality while improving spectral efficiency, digital transmission of TV signals offers an attractive replacement of traditional analog vestigial-sideband (VSB) modulation, as depicted in Fig. 2.36. Drawing on high-speed compression technology, image signals as well as other innovative services can be digitized and multiplexed into a single high-rate bit stream, then modulated to a standard 6-MHz channel allocation. With QAM techniques, this can provide a six-to-one improvement in bandwidth efficiency, suitable for use with older coaxial systems as well as newer fiber systems.

This type of digital broadcast dictates that blind equalization be a vital component of the subscriber's set-top receiver. The alternative, a conventional trained equalizer, requires valuable bandwidth to provide a retraining signal sufficiently often to facilitate appropriately rapid acquisition time. For example, an adequate response for the "channel surfing" subscriber might call for acquisition within a few hundred milliseconds of a tuning operation.

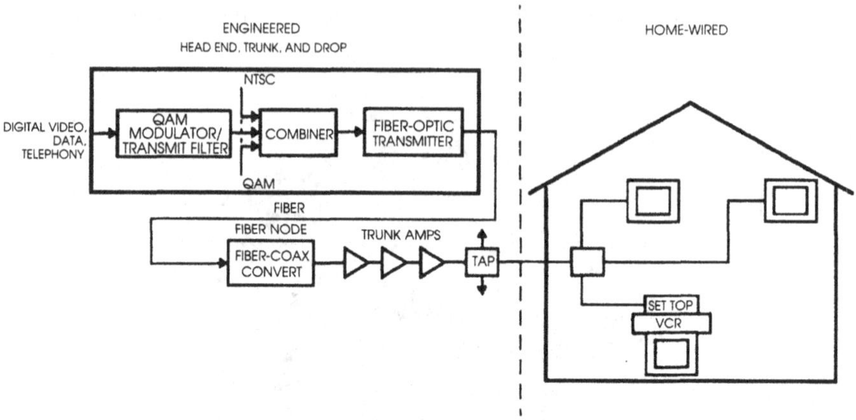

Figure 2.36 Digital cable television distribution.

Because the blind equalizer admits adaptation from just the broadcast data, no periodic transmission of a training signal is necessary. The receiver response to retuning is then basically limited by CM equalizer agility, which, as noted in Section 2.4, is determined by the choice of step size and equalizer length. Our goal in this section is to apply the rules for blind FSE design to this class of cable channels.

To characterize the nature of the cable channel, tests were conducted using typical residential segments; models of the dispersive and interference effects were obtained after exciting selected bands with digital test signals. Throughout this discussion, we refer to the cable-channel time response provided by the SPIB dataset, entitled `chan1.mat`. Analysis of an actual test signal provides this channel's effective response, as shown in Fig. 2.37. The magnitude of the complex-valued time response is shown in Fig. 2.37*b*, revealing modest time dispersion (i.e., microreflections) with an essentially symmetric distribution, and with the bulk of the energy limited to around ± 0.5 μs about its peak. In Fig. 2.37*a*, we see its frequency-response behavior: channel passband flat to within a couple of dB, as is typical of properly maintained cable systems.

In the current application, the TV signal is a 64-QAM symbol stream with a 5-MHz symbol rate, thus confined to the required 6-MHz allocation. The $T/2$ sampling rate of 10 MHz implies that the 1-μs dispersion noted in Fig. 2.37*b* spans about 10-tap intervals of the fractionally spaced equalizer. Due to the nature of the cable channel, the noise environment can be assumed to be benign with SNR in excess of 30 dB.

Using this scenario as representative for the cable TV environment, we consider the requirements for the receiver's equalizer implementation, that is, adequate asymptotic error performance and sufficient

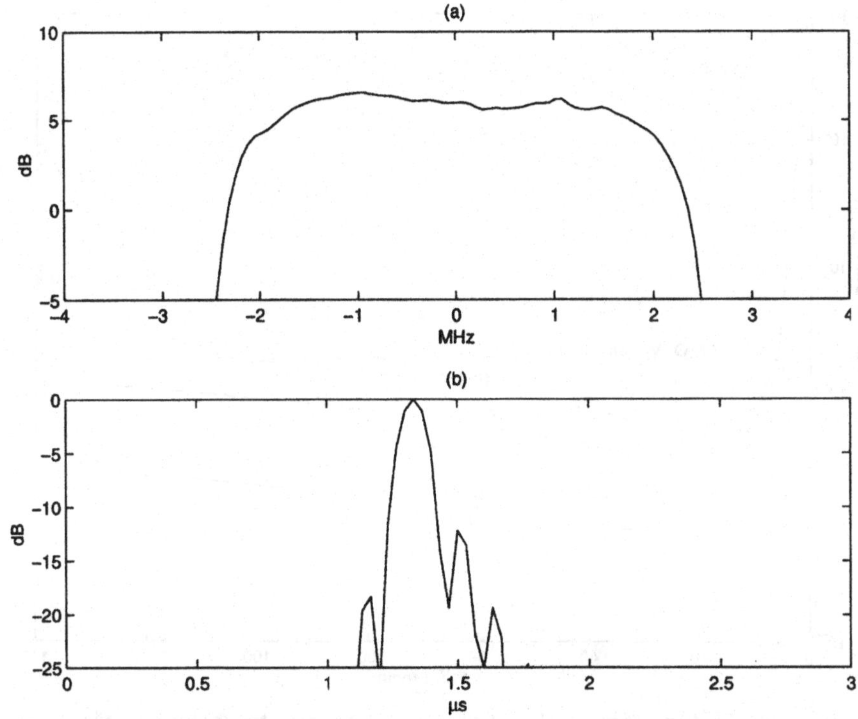

Figure 2.37 Prototypical cable TV channel: (*a*) frequency-response magnitude; (*b*) impulse-response magnitude.

agility. Drawing on the CMA undermodeling analysis and excess MSE analysis presented in Section 2.3, Fig. 2.38 depicts the estimated MSE performance of the CM receiver as a function of equalizer length, parameterized by three values of step size. The required MSE for reliable switchover to DD mode given the 64-QAM signal constellation is shown as a dashed line. A step-size selection of $\mu = 5 \times 10^{-4}$ provides a time constant on the order of a millisecond, presumed adequate in our design. Finally, choosing an equalizer length of 22, we expect an MSE sufficiently low (upon transition to the DD mode) to ensure convergence.

It was noted in Section 2.4 that proper initialization of CMA is essential to assure appropriate convergence. When initializing the FSE coefficients, the position of the impulsive double tap must be specified. Given a typical channel response, selection of the spike location is aided by the MMSE-versus-delay plot that can be calculated for a particular combination of equalizer length and SNR, as shown in Fig. 2.39*b*. As discussed in Section 2.4, the double-spike location is closely related to the system delay achieved by CMA (and likewise to the choice of

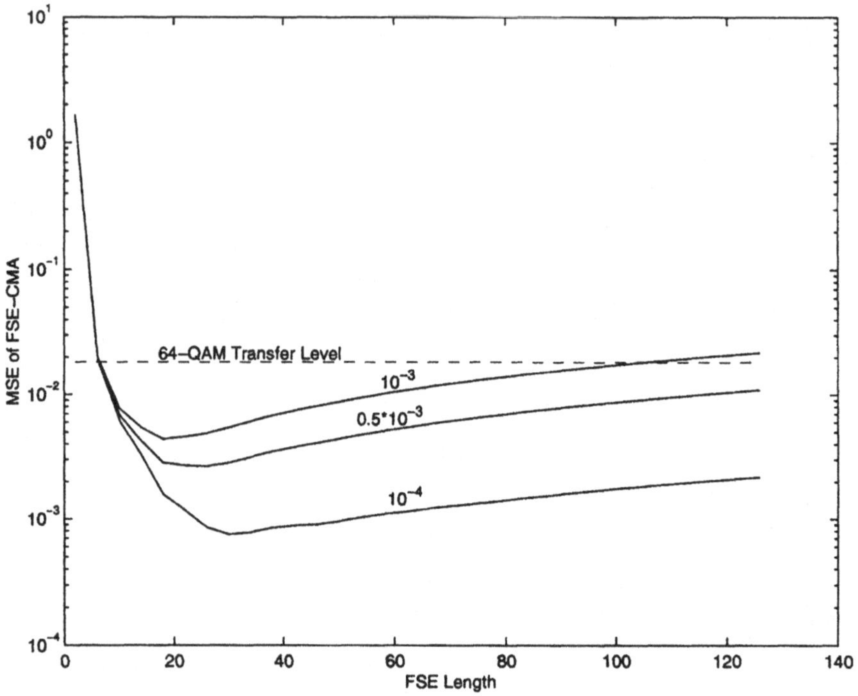

Figure 2.38 Prototypical cable TV channel: MSE due to undermodeling and excess MSE.

training sequence delay in the trained-LMS scenario). For our example, the symmetry of the channel's time response leads to a fairly constant minimum, symmetric about the center, spanning ± 5-baud intervals. The symmetric nature of the typical cable channel indicates that a double-spike initialization at taps 11 and 12 should be suitable in most situations.

To validate our design choices, Fig. 2.40 shows an ensemble-averaged MSE trajectory characterizing CMA adaptation. Ten sequences of 64-QAM symbols were passed through our typical channel model and the result applied to the CMA-adapted blind FSE. The plot shows transient behavior of the average MSE, where the acceptable "open-eye" condition necessary for switchover to DD adaptation occurs in about a millisecond.

Based on these observations, we can see that reliable blind reception of digital television signals is possible with reasonably simple and low-cost hardware at the receiver, as is critical for commercial acceptance. Note that our design approach has been purposely conservative in order

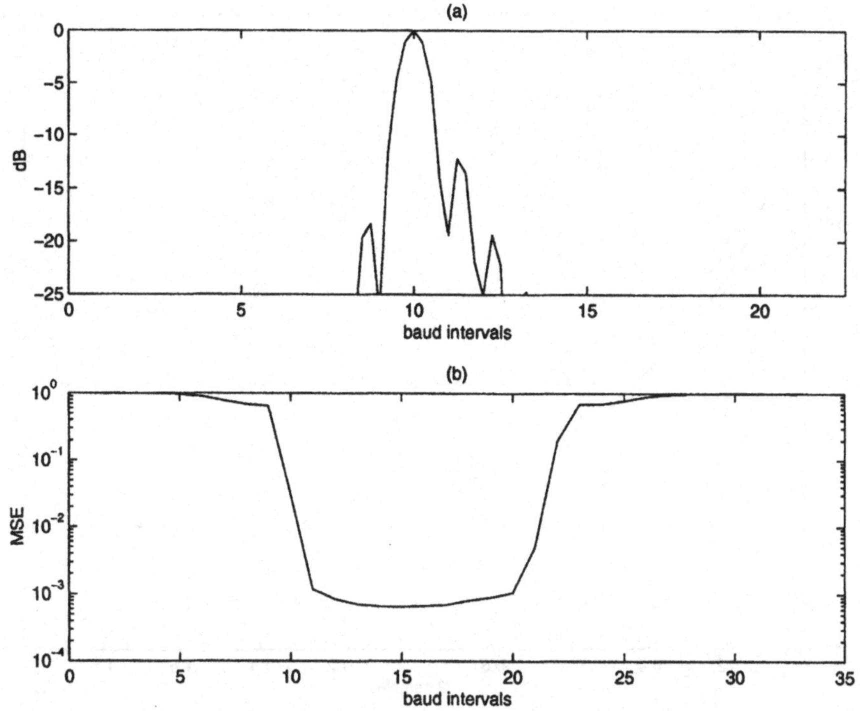

Figure 2.39 Sensitivity of equalizer initialization for a prototypical cable TV channel: (*a*) channel impulse-response magnitude (as in Fig. 2.37); (*b*) MMSE versus system delay for 22-tap *T*/2 FSE and 30-dB SNR.

to accommodate extremes in channel behavior that might be encountered in less ideal situations.

2.5.3 Mobile Radio: The Terrestrial Microwave Channel

In recent years, point-to-point microwave transmission, which traditionally used analog frequency modulation, has been aggressively converted to higher-efficiency digital-modulation techniques. This transition has been motivated by increased demands for bandwidth, and has been made possible by advances in high-speed processing technology over the last decade. A traditional point-to-point link deployment involves fixed-antenna installations with narrow collinear beams, carefully adjusted to minimize reflective interference. This type of system has traditionally served the needs of remote long-haul telephone service as well as intra-urban private voice, video, and data feeds. In this section we consider

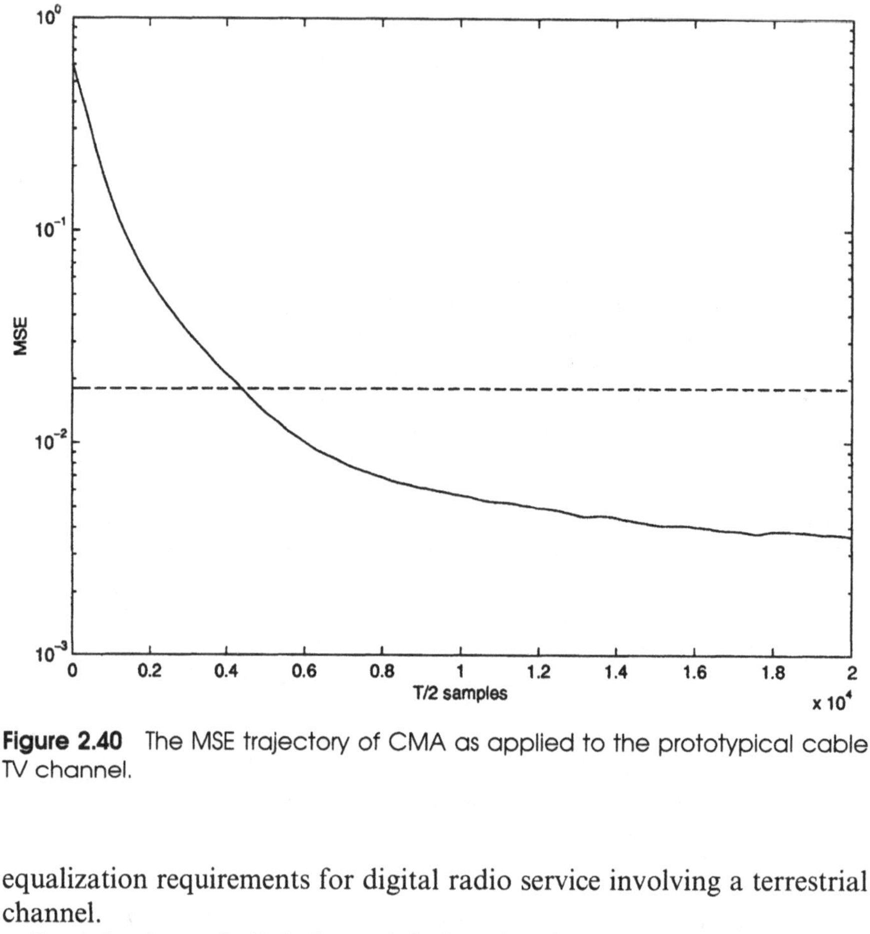

Figure 2.40 The MSE trajectory of CMA as applied to the prototypical cable TV channel.

equalization requirements for digital radio service involving a terrestrial channel.

Introduction of digital modulation in the microwave band has brought new exacting requirements for mitigation of propagation and interference effects. The characteristics of the point-to-point channel tend to be slowly varying, for example, broad attenuation due to atmospheric and meteorological conditions as well as frequency-selective losses resulting from reflective energy. Such multipath reflections result in correlated interference that generates a frequency-dependent passband ripple, that is, alternating constructive and destructive components. The period of the rippling is a function of differential path delay, and the amplitude is dependent on reflective attenuation. Figure 2.41 shows an actual radio-frequency (RF) channel subject to significant multipath; note how the delayed arrivals visible in the time domain result in a serious rippling of the passband magnitude.

The multipath environment is often a major factor in antenna design and placement, with the goal of minimizing the reflections from build-

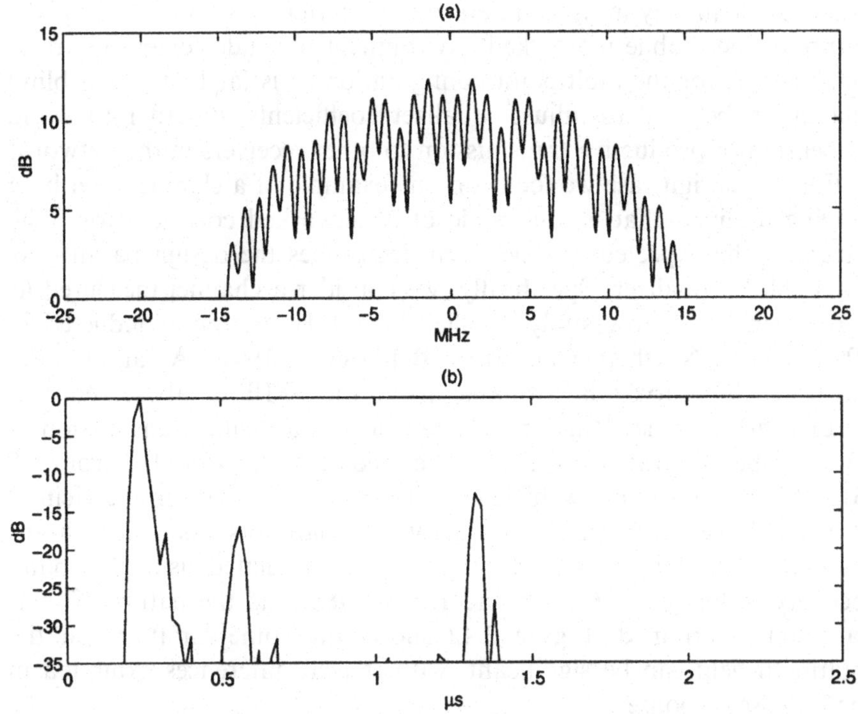

Figure 2.41 Terrestrial microwave channel: (*a*) frequency-response magnitude; (*b*) impulse-response magnitude.

ings and ground-surface features illuminated by the main beam. With the introduction of digital-modulation schemes using large-signal constellations, the use of active digital processing for equalization and polarization restoration has become a necessity. In the point-to-point communication environment, blind adaptation can be performed as a step in outage recovery, with switchover to decision-directed adaptation conducted during information exchange. Such an implementation is adequate to combat the normal slow channel variations and can often handle faster contributions due to dynamic reflectors, for example, vehicles and aircraft.

The situation is made considerably more complex with relative motion of receiver and transmitter. In a dynamic scenario, for example, a mobile network, vehicle motion, coupled with less directive RF transmission, induces variations in the reflective field responsible for interferometric behavior in the channel response (i.e., an in-band "wobbling" fade). This is further complicated by other distributed and motion-induced reflections, such as water motion or shimmering foliage. These features call for significant improvement in receiver tracking ability in order to adequately mitigate intersymbol interference. Furthermore, the nature of the mobile networked environment precludes extensive training. Instead, for the multipoint configuration, it is far better that blind adaptation be used to adjust equalizer coefficients, in order to avoid suspension of productive transmission to other receivers in the network.

For this design case we focus on one example of a channel seen by a moving receiver, that is, one node of a networked configuration. The nature of the expected channel then determines the design parameters of the blind equalizer. Specifically, we consider a channel measured as part of an RF survey study Applied Signal Technology conducted in 1996 in rural Northern California (Behm et al. 1997). A sample data set from this survey is available from the SPIB database, entitled `hillshadow.otm.4GHz.snp`. Snapshots of a digital transmission at 25 Msymbol/s, downconverted to intermediate frequency (IF) from 4.5 GHz, were taken from a vehicle traveling at modest speed in the vicinity of a fixed transmitter site. By analyzing snapshots taken at 200-ms intervals, the channel characteristics can be estimated using the blind recovery technique of Gooch and Harp (1988) and the nature of their variation determined. Figure 2.42 shows that time-variations in the multipath field can be significant, with radical differences exhibited in the impulse response.

This case represents conflicting design goals: the channel delay spread exhibited in Fig. 2.41*b*, typically spanning 10 to 25 symbols (400 to 1000 ns), calls for a relatively long equalizer for adequate ISI recovery, and thus to maintain suitable steady-state performance, the step size must be kept small. However, the time-varying nature of the mobile channel requires a step size large enough to track the typical multipath variations exhibited by Fig. 2.42.

In Sidebar C of Treichler et al. (1996), the minimum recommended $T/2$-spaced equalizer length for the QPSK source used here is 2 taps for each symbol period of delay spread. With a noted maximum delay spread of approximately 25 symbols, and using an additional factor-of-2 margin-of-safety, we choose 100 equalizer taps.

From Sidebar D of Treichler et al. (1996), with a unit-variance received signal and an equalizer of up to 100 taps, the recommended upper bound on the step size (of LMS[21]) is 2 divided by the number of taps,

[21] In our application of LMS step-size and length selection guidelines to CMA, we exploit the similarities between LMS and CMA convergence behavior discussed in Section 2.4 (and more technically in Section 2.3). Though CMA and LMS convergence rates may differ by a factor of 2 or so, we regard such differences to be subsumed by our wide safety margins.

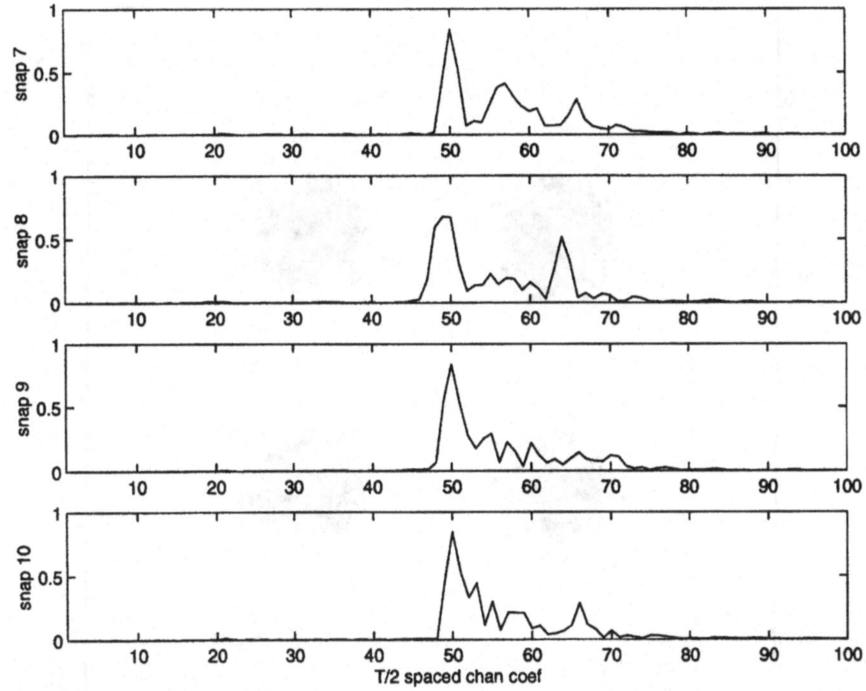

Figure 2.42 Mobile microwave: estimates of channel impulse responses derived from data records separated by 200-ms intervals.

that is, 0.02 in our design. The minimum step size is related to the variability of the channels of interest. For example, Fig. 2.42 exhibits substantial variation every 200 ms. Presuming that we wish to have a convergence time constant less than 100 times this 0.2-s interval, the average time constant formula for LMS suggests [as discussed in Sidebar D of Treichler et al. (1996)] that for a $T/2$-spaced FSE,

$$\mu > \frac{1}{(\text{no. of taps}) \cdot (\text{no. of baud intervals per second}) \cdot (\text{minimum convergence-time constant})}$$

For 25 Mbaud/s and 100 equalizer taps, this implies that μ should be chosen greater than 2×10^{-5}. Hence, we choose $\mu = 10^{-3}$.

Our choices of length and step size allowed us to equalize and accurately demodulate the QPSK signal corresponding to the channel shown in Fig. 2.42. This is demonstrated by the equalized constellation in Fig. 2.43 for the source constellation of Fig. 2.44.

This discussion has illustrated the use of blind techniques for digital transmission over three very different channels. In each case, the moti-

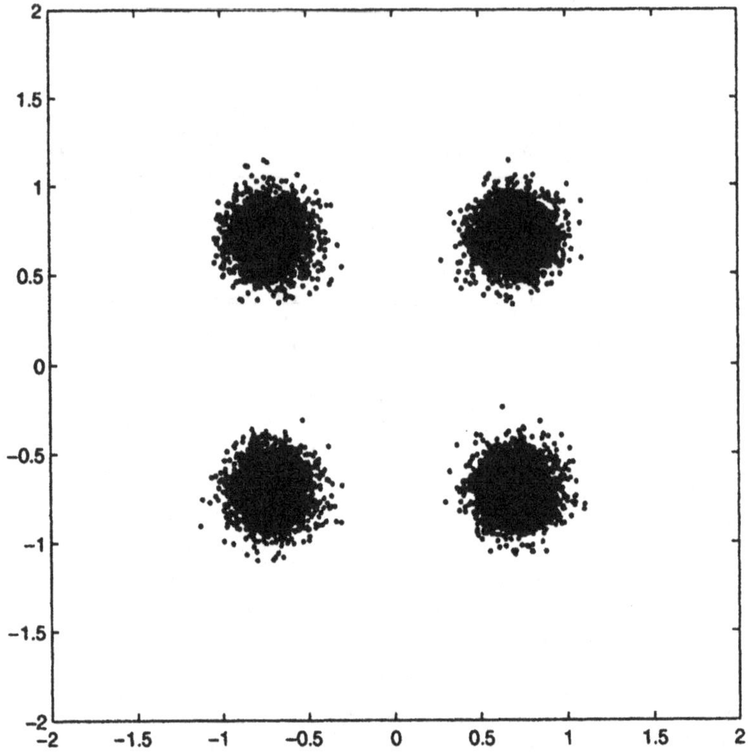

Figure 2.43 Equalized constellation resulting from the design parameters chosen in the mobile radio example, exhibiting a cluster variance of −16.3 dB.

vation is blind adaptation of the equalizer to avoid disruption of the transmitter's information transfer. And in each case, once adjustment is sufficient to restore constellation integrity, the conventional DD mode is begun. One should note that, for each scenario, future applications involving more elaborate network topologies hinge on the success of blind equalization.

2.6 CONCLUSIONS

The most basic of adaptive filter configurations in practice and pedagogy is a tapped delay line with impulse-response coefficients updated via the LMS algorithm in the presence of a replica of the desired filter output sequence (e.g., a training sequence). Thirty years of experience have taught us how to use LMS even though the theory describing its behavior in all pragmatic circumstances is still not complete. We are in an

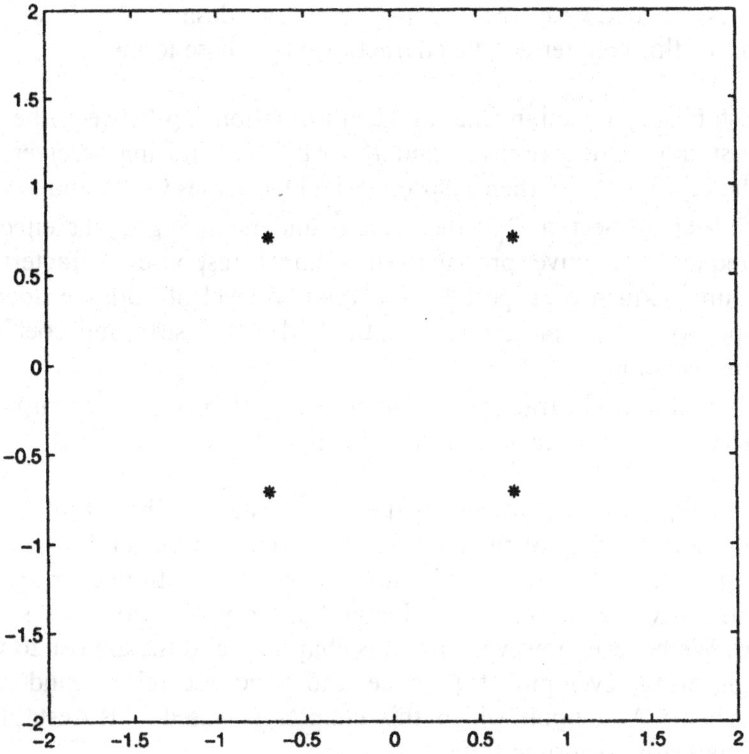

Figure 2.44 QPSK constellation used in the mobile radio example.

earlier period in a similar pattern of growing utilization and understanding of the CMA, which has proven to be the baseline algorithm for a variety of applications where the replica of the desired filter output is unavailable, that is, the blind scenario.

The movement from point-to-point, session-oriented communications, for which cooperative equalizer training was acceptable, to digital multipoint and broadcast systems, for which training is unacceptable, mandates blind equalization. Because CMA is the one blind algorithm for adaptive equalization that has seen widespread success in operational communication systems, we have assembled a set of design guidelines for it in this chapter.

The basic pedagogy employed in this chapter for organizing a behavior theory for a blind fractionally spaced equalizer is similar to that used in a variety of widely cited textbooks for describing trained, baud-spaced equalizer behavior theory. This approach is bolstered by recent results—described in previous sections of this chapter—on the similarity of CM and MSE cost-function minima. This prompts the adoption of the taxonomy associated with LMS-adapted equalizer design rules. Indeed, the content of this chapter can be characterized in these terms:

- Establish (in Sections 2.2 and 2.3) a relationship between the CM-cost minimizing receiver and the MSE minimizing receiver, and therefore between their respective SGD schemes CMA and LMS.
- Collect (in Section 2.3) the current understanding of the effects of frequency-selective propagation channel response characteristics, source-sequence properties, and relative level of additive noise on appropriate adaptive filter length, CMA step size, and coefficient initialization.
- Extract and illustrate design guidelines with academic examples (in Section 2.4) and case studies (in Section 2.5).

This chapter, as a chronicle of the current state of the art in converting an understanding of the behavior of CMA into useful design guidelines, indicates that there is still much more to do, both in terms of fundamental advances in understanding and in terms of a variety of specific details. We believe, however, that this chapter—and the almost 20 years of engineering development, practice, and basic research in blind equalization via CMA upon which this chapter is based—also establishes CMA as ready for prime time.

REFERENCES

Axford, Jr., R. A., L. B. Milstein, and J. R. Zeidler, Feb. 1998, "The effects on PN sequences on the misconvergence of the constant modulus algorithm," *IEEE Trans. Signal Processing*, vol. 46, no. 2, pp. 519–23.

Behm, J. D., T. J. Endres, P. Schniter, C. R. Johnson, Jr., C. Prettie, M. L. Alberi, and I. Fijalkow, 1997, "Characterization of an empirically-derived database of time-varying microwave channel responses," *Proc. Asilomar Conference on Signals, Systems and Computers* (Pacific Grove, CA), pp. 1549–1553, 2–5 Nov. 1997.

Bellini, S., 1994, "Bussgang techniques for blind deconvolution and equalization," in S. Haykin, ed., *Blind Deconvolution* (Englewood Cliffs, NJ: Prentice-Hall) pp. 60–120.

Benveniste, A., M. M'etivier, and P. Priouret, 1990, *Adaptive Algorithms and Stochastic Approximations* (Paris, France: Springer-Verlag).

Bussgang, J. J., 1952, *Crosscorrelation Functions of Amplitude-Distorted Gaussian Signals*, Technical Report 216, MIT Research Laboratory of Electronics, Cambridge, MA.

Caines, P. E., 1988, *Linear Stochastic Systems* (New York: John Wiley).

Casas, R. A., F. Lopez de Victoria, I. Fijalkow, P. Schniter, T. J. Endres, and C. R. Johnson, Jr., 1997, "On MMSE fractionally-spaced equalizer design," *Proc. Int. Conf. Digital Signal Processing* (Santorini, Greece), pp. 395–398, 2–4 July 1997.

Chung, W., and C. R. Johnson, Jr., 1998, "Characterization of the regions of convergence of CMA adaptive blind fractionally spaced equalizers," to appear in *Proc. Asilomar Conference on Signals, Systems and Computers* (Pacific Grove, CA), 2–4 Nov. 1998.

Chung, W., and J. P. LeBlanc, 1998, "The local minima of fractionally-spaced CMA blind equalizer cost function in the presence of channel noise," *Proc. IEEE Int. Conf. Acoustics, Speech, and Signal Processing* (Seattle, WA), pp. 3345–3348, 12–15 May 1998.

Cioffi, J. M., G. P. Dudevoir, M. V. Eyuboglu, and G. D. Forney, Jr., Oct. 1995, "MMSE decision feedback equalization and coding—Part I and II," *IEEE Trans. Communic.* vol. 43, no. 10, pp. 2582–2604.

Ding, Z., R. A. Kennedy, B. D. O. Anderson, and C. R. Johnson, Jr., Sep. 1991, "Ill-convergence of Godard blind equalizers in data communication systems," *IEEE Trans. Communic.*, vol. 39, no. 9, pp. 1313–1327.

Endres, T. J., B. D. O. Anderson, C. R. Johnson, Jr., and M. Green, 1997a, "On the robustness of the fractionally-spaced constant modulus criterion to channel order undermodeling: Part I," *Proc. IEEE Signal Processing Workshop on Signal Processing Advances in Wireless Communic.* (Paris, France), pp. 37–40, 16–18 Apr. 1997.

Endres, T. J., B. D. O. Anderson, C. R. Johnson, Jr., and M. Green, 1997b, "On the robustness of the fractionally-spaced constant modulus criterion to channel order undermodeling: Part II," *Proc. IEEE Int. Conf. on Acoustics, Speech, and Signal Processing* (Munich, Germany), pp. 3605–3608, 20–24 Apr. 1997.

Ericson, T., May 1971, "Structure of optimum receiving filters in data transmission systems," *IEEE Trans. Inform. Theory*, vol. 17, no. 3, pp. 352–353.

Fijalkow, I., 1996, "Multichannel equalization lower bound: A function of channel noise and disparity," *Proc. IEEE Signal Processing Workshop on Statistical Signal and Array Processing* (Corfu, Greece), pp. 344–347, June 24–26, 1996.

Fijalkow, I., F. Lopez de Victoria, and C. R. Johnson, Jr., 1994, "Adaptive fractionally spaced blind equalization," *Proc. IEEE Signal Processing Workshop* (Yosemite National Park, CA), pp. 257–260, 2–5 Oct. 1994.

Fijalkow, I., C. Manlove, and C. R. Johnson, Jr., Jan. 1998, "Adaptive fractionally spaced blind CMA equalization: Excess MSE," *IEEE Trans. Signal Processing*, vol. 46, no. 1, pp. 227–231.

Fijalkow, I., A. Touzni and C. R. Johnson Jr., Sep. 1995, "Spatio-temporal equalizability under channel noise and loss of disparity," *Proc. Colloque*

GRETSI sur le Traitement du Signal et des Images (Sophia Antipolis, France), pp. 293–296.

Fijalkow, I., A. Touzni, and J. R. Treichler, Jan. 1997, "Fractionally spaced equalization using CMA: Robustness to channel noise and lack of disparity," *IEEE Trans. Signal Processing*, vol. 45, no. 1, pp. 56–66.

Foschini, G. J., Oct. 1985, "Equalizing without altering or detecting data (digital radio systems)," *AT&T Tech. J.*, vol. 64, no. 8, pp. 1885–1911.

Fuhrmann, P. A., 1996, *A Polynomial Approach to Linear Algebra* (New York: Springer-Verlag).

Gitlin, R. D., J. F. Hayes, and S. B. Weinstein, 1992, *Data Communications Principles* (New York: Plenum Press).

Godard, D. N., Nov. 1980, "Self-recovering equalization and carrier tracking in two-dimensional data communication systems," *IEEE Trans. Communic.*, vol. 28, no. 11, pp. 1867–1875.

Gooch, R. P., and J. C. Harp, 1988, "Blind channel identification using the constant modulus adaptive algorithm," *Proc. IEEE Int. Conf. Communic.* (Philadelphia, PA), pp. 75–79, June 12–15, 1988.

Gray, R. M., Nov. 1980, "On the asymptotic eigenvalue distribution of Toeplitz matrices," *IEEE Trans. Inform. Theory*, vol. 23, pp. 357–368.

Gu, M. and L. Tong, Oct. 1999, "Geometrical characterizations of constant modulus receivers," *IEEE Trans. on Signal Processing*, vol. 47, pp. 2745–2756.

Haykin, S., ed., 1994, *Blind Deconvolution* (Englewood Cliffs, NJ: Prentice-Hall).

Haykin, S., 1996, *Adaptive Filter Theory*, 3rd ed., (Englewood Cliffs, NJ: Prentice-Hall).

International Telecommunication Union, 1988, *Data communication over the telephone network: 9600 bits per second modem standardized for use on point-to-point 4-wire leased telephone-type circuits*, ITU-T Recommendation V.29, Blue Book, vol. VIII.I, Geneva, Switzerland.

Johnson, Jr., C. R., and B. D. O. Anderson, July–Aug. 1995, "Godard blind equalizer error surface characteristics: White, zero-mean, binary case," *Int. J. Adaptive Control and Signal Processing*, vol. 9, pp. 301–324.

Johnson, Jr., C. R., P. Schniter, T. J. Endres, J. D. Behm, D. R. Brown, and R. A. Casas, Oct. 1998, "Blind equalization using the constant modulus criterion: A review," *Proc. IEEE* (*Special issue on Blind System Identification and Estimation*), vol. 86, no. 10, pp. 1927–1950.

Kailath, T., 1980, *Linear Systems* (Englewood Cliffs, NJ: Prentice-Hall).

Lambotharan, S., J. Chambers, and C. R. Johnson, Jr., June 1997, "Attraction of saddles and slow convergence in CMA adaptation," *Signal Processing*, vol. 59, no. 2, pp. 335–340.

Larimore, M. G., and J. R. Treichler, 1983, "Convergence behavior of the constant modulus algorithm," *Proc. IEEE Int. Conf. on Acoustics, Speech, and Signal Processing* (Boston, MA), pp. 13–16, 14–16 Apr. 1983.

Larimore, M. G., S. L. Wood, and J. R. Treichler, 1997, "Performance costs for theoretical minimum-length equalizers," *Proc. IEEE Int. Conf. on Acoustics, Speech, and Signal Processing* (Munich, Germany), pp. 2477–2480, 21–24 Apr. 1997.

LeBlanc, J. P., Aug. 1995, Effects of source distributions and correlation on fractionally spaced blind constant modulus algorithm equalizers, Ph.D. dissertation, Cornell University, Ithaca, NY.

LeBlanc, J. P., I. Fijalkow, and C. R. Johnson, Jr., Mar. 1998, "CMA fractionally spaced equalizers: Stationary points and stability under IID and temporally correlated sources," *Int. J. Adaptive Control and Signal Processing*, vol. 12, no. 2, pp. 135–155.

Lee, E. A., and D. G. Messerschmitt, 1994, *Digital Communication*, 2nd ed. (Boston, MA: Kluwer Academic Publishers).

Li, Y., and Z. Ding, Sep. 1995, "Convergence analysis of finite length blind adaptive equalizers," *IEEE Trans. on Signal Processing*, vol. 43, no. 9, pp. 2120–2129.

Li, Y., and Z. Ding, Apr. 1996, "Global convergence of fractionally spaced Godard (CMA) adaptive equalizers," *IEEE Trans. Signal Processing*, vol. 44, no. 4, pp. 818–826.

Li, Y., and K. J. R. Liu, Nov. 1996, "Static and dynamic convergence behavior of adaptive blind equalizers," *IEEE Trans. Signal Processing*, vol. 44, no. 11, pp. 2736–2745.

Li, Y., K. J. R. Liu, and Z. Ding, Nov. 1996, "Length and cost dependent local minima of unconstrained blind channel equalizers," *IEEE Trans. on Signal Processing*, vol. 44, no. 11, pp. 2726–2735.

Liu, R. W., ed., Oct. 1998, *Proc. IEEE: Special Issue on Blind System Identification and Estimation*, vol. 86, no. 10 (Piscataway, NJ: Institute of Electrical and Electronic Engineers).

Ljung, L., and T. Soderstrom, 1983, *Theory and Practice of Recursive Identification* (Cambridge, MA: MIT Press).

Lucky, R. W., Feb. 1966, "Techniques for adaptive equalization of digital communication systems," *Bell System Tech. J.*, vol. 45, pp. 255–286.

Luenberger, D. G., 1990, *Optimization by Vector Space Methods*, 2nd ed. (New York: Wiley).

Macchi, O., 1995, *Adaptive Processing: The Least Mean Squares Approach with Applications in Transmission* (Chichester, NY: Wiley).

Moulines, E., P. Duhamel, J. Cardoso, and S. Mayrargue, Feb. 1995, "Subspace methods for blind identification of multichannel FIR filters," *IEEE Trans. Signal Processing*, vol. 43, no. 2, pp. 516–525.

Papadias, C. B., 1997, "On the existence of undesired global minima of Godard equalizers," *Proc. IEEE International Conference on Acoustics, Speech, and Signal Processing* (Munich, Germany), pp. 3937–3940, 20–24 Apr. 1997.

Paulraj, A. J., and C. B. Papadias, Nov. 1997, "Space-time processing for wireless communications," *IEEE Signal Processing Mag.*, vol. 14, no. 6, pp. 49–83.

Proakis, J. G., 1995, *Digital Communications*, 3rd ed. (New York: McGraw-Hill).

Qureshi, S. U. H., Sep. 1985, "Adaptive equalization," *Proc. IEEE*, vol. 73, no. 9, pp. 1349–1387.

Rosenblatt, M., 1985, *Stationary Sequences and Random Fields* (Boston, MA: Birkhäuser).

Strang, G., 1988, *Linear Algebra and its Applications*, 3rd ed. (Fort Worth, TX: Harcourt Brace Jovanovich).

Tong, L., and H. Zeng, Mar. 1997, "Channel surfing re-initialization for the constant modulus algorithm," *IEEE Signal Processing Lett.*, vol. 4, no. 3, pp. 85–87.

Tong, L., G. Xu, B. Hassibi, and T. Kailath, Jan. 1995, "Blind identification and equalization based on second-order statistics: A frequency-domain approach," *IEEE Trans. on Information Theory*, vol. 41, no. 1, pp. 329–334.

Touzni, A., and I. Fijalkow, Sept. 1996, "Does fractionally-spaced CMA converge faster than LMS?" *Proc. European Signal Processing Conference* (Trieste, Italy), pp. 1227–1230.

Treichler, J. R., and B. G. Agee, Apr. 1983, "A new approach to multipath correction of constant modulus signals," *IEEE Trans. Acoustics, Speech, and Signal Processing*, vol. ASSP-31, no. 2, pp. 459–472.

Treichler, J. R., I. Fijalkow, and C. R. Johnson, Jr., May 1996, "Fractionally-spaced equalizers: How long should they really be?" *IEEE Signal Processing Mag.*, vol. 13, no. 3, pp. 65–81.

Treichler, J. R., M. G. Larimore, and J. C. Harp, Oct. 1998, "Practical blind demodulators for high-order QAM signals," *Proc. IEEE: special issue on Blind System Identification and Estimation*, vol. 86, no. 10, pp. 1907–1926.

Ungerboeck, G., Aug. 1976, "Fractional tap-spacing equalizer and consequences for clock recovery in data modems," *IEEE Trans. Communic.*, vol. 24, no. 8, pp. 856–864.

Widrow, B., and S. D. Stearns, 1985, *Adaptive Signal Processing* (Englewood Cliffs, NJ: Prentice Hall).

Widrow, B., J. M. McCool, M. G. Larimore, and C. R. Johnson, Jr., Aug. 1976, "Stationary and non-stationary learning characteristics of the LMS adaptive filter," *Proc. IEEE*, vol. 64, no. 8, pp. 1151–1162.

Zeng, H. H., L. Tong, and C. R. Johnson, Jr., Nov. 1996a, "Behavior of fractionally-spaced constant modulus algorithm: Mean square error, robustness and local minima," *Proc. Asilomar Conference on Signals, Systems and Computers* (Pacific Grove, CA), pp. 305–309.

Zeng, S., H. H. Zeng, and L. Tong, June 1996b, "Blind equalization using CMA: Performance analysis and a new algorithm," *Proc. IEEE Int. Conf. Communic.* (Dallas, TX), pp. 847–851.

Zeng, H. H., L. Tong, and C. R. Johnson, Jr., July 1998, "Relationships between the constant modulus and Wiener receivers," *IEEE Trans. Inform. Theory*, vol. 44, no. 4, pp. 1523–1538.

3

RELATIONSHIPS BETWEEN BLIND DECONVOLUTION AND BLIND SOURCE SEPARATION

Scott C. Douglas and Simon Haykin

ABSTRACT

In this chapter, the relationships between the two related tasks of blind deconvolution and blind source separation are explored. The maximum-likelihood method for parameter estimation is shown to provide a unifying framework for deriving blind deconvolution and blind source-separation algorithms. To illustrate the structural relationships between the two tasks, the problem of blind source separation under circulant mixing conditions is considered, iterative algorithms for its solution are derived, and then these algorithms are related to recently proposed blind deconvolution techniques. The results of these various studies suggest the potential benefits that can be obtained from the cross-fertilization of these two fields.

3.1 INTRODUCTION

Blind source separation (BSS)—the task of forming multiple independent signals from sets of linear mixtures without the use of training

Unsupervised Adaptive Filtering, Volume II, Edited by Simon Haykin.
ISBN 0-471-37941-7 © 2000 John Wiley & Sons, Inc.

signals—is a problem that has garnered much recent research interest, as evidenced by the material presented in Chapters 2 to 9 of Volume I of this work. Driving these efforts are the many potential and varied applications of these methods, such as cochannel interference mitigation in wireless multiple-access communication systems (Cardoso and Laheld 1997; Honig et al. 1995; Paulraj and Papadias 1997) speech separation for improved audition in audio recording and transmission systems (Amari et al. 1997c; Lambert and Bell 1997; Torkkola 1996), and signal extraction for multisensor biological-signal recording systems (McKeown et al. 1998; Makeig et al. 1997), just to name a few. BSS is considered by many in the engineering and scientific disciplines to be a challenging task without an obvious formulation for its solution, factors that add to its research appeal.

By contrast, algorithms for blind deconvolution and equalization[1]—the task of forming a sequence of independent samples from a noisy filtered version of the sequence—have been in existence since the pioneering work of Lucky on decision-directed (DD) equalization at AT&T Bell Laboratories in the mid-1960s (Lucky 1966). Work in this field has clearly reached a substantial level of maturity, as many techniques, including DD equalization (Qureshi 1985; Sato 1975) and the Godard/constant-modulus algorithm (CMA) (Godard 1980; Treichler and Agee 1983) have been incorporated into successful products (Treichler et al. 1998). Although blind deconvolution and equalization continue to be interesting research topics, recent efforts are clearly more focused on the practical details surrounding particular algorithm implementations, as evidenced by the work on CMA described in the previous chapter.

An obvious question to ask, then, is: What makes BSS so much different from blind deconvolution? It is this question that we seek to address in this chapter. It is our hope that, by comparing and contrasting the two problem genres, the salient features of each will be illuminated clearly. In addition, we provide a formalism that allows the translation of any given source-separation algorithm to a blind deconvolution algorithm. Such concepts are quite closely related to the work of Lambert and Nikias as described in Lambert (1996) as well as in Chapter 9 of Volume I of this work, although the approach taken in this chapter is somewhat different and relies on connections between Toeplitz and circulant matrices as outlined by Gray (1977).

[1] Throughout our discussion, we shall use the words "deconvolution" and "equalization" interchangeably, although the latter term is generally restricted to problems involving discrete-amplitude source signals, for example, as typically found in digital communication tasks.

In our discussion, we follow the notational conventions largely adopted by other chapter authors in this book. Listed in the following Notation section are the variable and function names and symbols that we employ in our discussions. In addition, all signals are assumed to be either strict-sense stationary random processes or realizations of such random processes, as inferred from context.

3.1.1 Notation

VARIABLE DEFINITIONS FOR SOURCE SEPARATION

Variable	Description	Dimension
k	Time index	(1×1)
m	Number of sources	(1×1)
n	Number of measured signals	(1×1)
$\mathbf{s}(k)$	Source-signal vector	$(m \times 1)$
$s_i(k)$	ith source signal	(1×1)
$\mathbf{v}(k)$	Noise vector	$(n \times 1)$
$v_i(k)$	ith noise signal	(1×1)
$\mathbf{x}(k)$	Received-signal vector	$(n \times 1)$
$x_i(k)$	ith received signal	(1×1)
$\mathbf{y}(k)$	Output-signal vector	$(m \times 1)$
$y_i(k)$	ith output signal	(1×1)
$\mathbf{A}$	Unknown channel matrix	$(n \times m)$
$\mathbf{W}(k)$	Separating matrix	$(m \times n)$
$\mathbf{P}$	Permutation matrix	$(m \times m)$
$\mathbf{D}$	Diagonal scaling matrix	$(m \times m)$
$\mathbf{C}$	Combined system matrix	$(m \times m)$
$\mathbf{R_{sx}}$	Source-input cross-correlation matrix	$(m \times n)$
$\mathbf{R_{xx}}$	Input signal autocorrelation matrix	$(n \times n)$
σ_v^2	Variance of $v_i(k)$	(1×1)
$f_{\mathbf{y}}(\mathbf{y})$	Joint pdf of $\mathbf{y}(k)$	(1×1)
$f_i(y_i)$	Marginal pdf of $y_i(k)$	(1×1)

VARIABLE DEFINITIONS FOR DECONVOLUTION

k	Time index
z^{-1}	Unit delay
$s(k)$	Source signal

$v(k)$	Noise signal
$x(k)$	Received signal
$X(z)$	z-transform of received signal
$y(k)$	Output signal
$Y(z)$	z-transform of output signal
a_j	Channel impulse response
$A(z)$	Channel system function
w_j	Equalizer impulse response
$W(z)$	Equalizer system function
c	Channel/equalizer gain
Δ	Channel/equalizer group delay
$C(z)$	Combined system function
$r_{xx}(j)$	Input-signal autocorrelation function
σ_v^2	Variance of $v(k)$

VARIABLE DEFINITIONS FOR ALGORITHMIC RELATIONSHIPS

$\mathbf{W}_{\mathrm{ZF}}$ or $W_{\mathrm{ZF}}(z)$	Zero-forcing solution
$\mathcal{J}_{\mathrm{MSE}}(\mathbf{W})$ or $\mathcal{J}_{\mathrm{MSE}}(W(z))$	MSE cost function
$\mathbf{W}_{\mathrm{MSE}}$ or $W_{\mathrm{MSE}}(z)$	Minimum MSE solution
$\hat{\mathbf{A}}_{\mathrm{LS}}$	Least-squares channel estimate
$\hat{\mathbf{W}}_{\mathrm{LS}}$	Least-squares channel inverse estimate
$\hat{\mathbf{R}}_{\mathbf{sx}}$	Source-input cross-correlation matrix estimate
$\hat{\mathbf{R}}_{\mathbf{ss}}$	Source-signal autocorrelation matrix estimate
$\hat{\mathbf{R}}_{\mathbf{xx}}$	Input-signal autocorrelation matrix estimate
$\mathbf{v}$	Generic random vector
θ	Generic parameter vector
$\hat{f}_{\mathbf{v}}(\mathbf{v}, \theta)$	Estimated distribution of $\mathbf{v}$
$f_{\mathbf{v}}(\mathbf{v}, \theta)$	True distribution of $\mathbf{v}$
$\mathcal{J}(\mathbf{W})$	Cost function
$H(f_{\mathbf{v}})$	Entropy of $\mathbf{v}$
$K(f_{\mathbf{v}} \| \hat{f}_{\mathbf{v}})$	Kullback-Leibler divergence
$\hat{\mathcal{J}}(\theta)$	Instantaneous cost function of θ
$\mu(k)$	Step-size sequence
$\phi_i(y)$	Score function of $y_i(k)$

VARIABLE DEFINITIONS FOR STRUCTURAL RELATIONSHIPS

M	Channel-length parameter
L	Equalizer-length parameter
$\mathbf{I}_j$	jth lower-diagonal circulant identity matrix
$\hat{\mathcal{J}}(\mathbf{W})$	Instantaneous cost function of $\mathbf{W}$
$\hat{\mathbf{g}}(\mathbf{y}(k))$	Estimated score-function vector
$\mathbf{u}(k)$	Auxiliary output-signal vector
$u_i(k)$	ith auxiliary output signal
$\mathcal{H}$	DFT of channel impulse response
$\mathcal{W}$	DFT of equalizer impulse response
$\mathcal{G}(k)$	DFT of estimated score-function vector
$\mathcal{U}(k)$	DFT of auxiliary output-signal vector
$\mathbf{1}$	Ones vector
K and α	Distribution constants
σ_y^2	Output-signal variance
σ_s^2	Source-signal variance
$\mathbf{A}_i$	ith block-circulant channel matrix

FUNCTION DEFINITIONS

$\mathbf{M}^*$	Element-by-element conjugate of $\mathbf{M}$
$\mathbf{M}^T$	Transpose of $\mathbf{M}$
$\mathbf{M}^\dagger$	Hermitian transpose of $\mathbf{M}$
$\mathbf{M}^{-1}$	Inverse of $\mathbf{M}$
$\mathbf{M}^+$	Pseudoinverse of $\mathbf{M}$
$\mathrm{tr}_j[\mathbf{M}]$	jth lower-diagonal circulant trace of $\mathbf{M}$
$[i]_n$	Modulo-n operation of i
$\odot$	Point-by-point vector multiplication
$\oslash$	Point-by-point vector division
$E\{\mathbf{v}\}$	Expectation of $\mathbf{v}$
$\partial f(\mathbf{M})/\partial \mathbf{M}$	Matrix of partial derivatives $\partial f(\mathbf{M})/\partial m_{ij}$

3.2 PROBLEM DESCRIPTIONS

3.2.1 Source Separation

We begin our discussion by briefly reviewing the mathematical and statistical formulation of the source-separation task (Cardoso 1998; Comon 1994). Figure 3.1a shows this task's structure, in which an n-dimensional

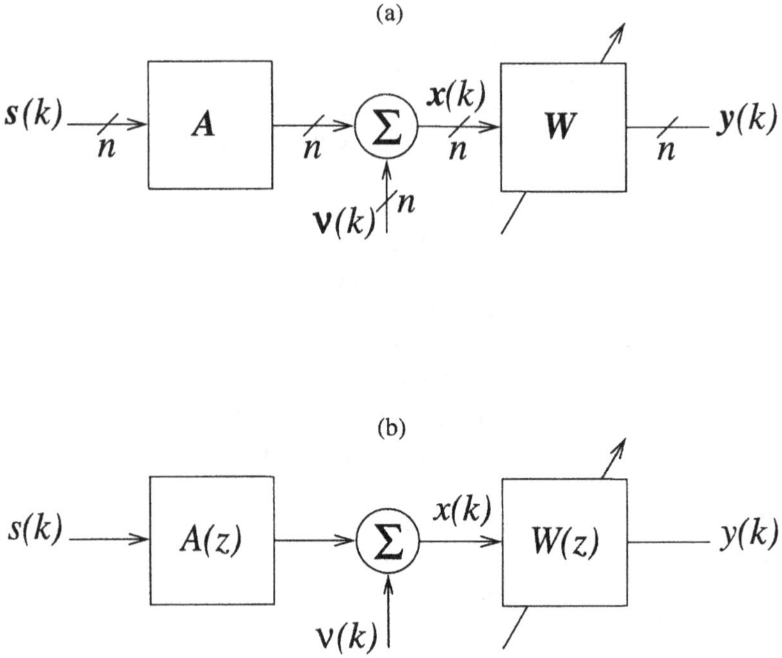

Figure 3.1 Block diagrams for the (*a*) source-separation, and (*b*) deconvolution tasks.

vector $\mathbf{x}(k) = [x_0(k) \;\; x_1(k) \cdots x_{n-1}(k)]^T$ of received signals at time k is produced from an m-dimensional vector $\mathbf{s}(k) = [s_0(k) \;\; s_1(k) \cdots s_{m-1}(k)]^T$ of source signals via the linear model

$$\mathbf{x}(k) = \mathbf{A}\mathbf{s}(k) + \boldsymbol{\nu}(k) \tag{3.1}$$

where $\mathbf{A}$ is an $(n \times m)$ channel matrix and $\boldsymbol{\nu}(k) = [\nu_0(k) \cdots \nu_{n-1}(k)]^T$ is a vector of noise components. The task at hand is to process the measurement vectors $\mathbf{x}(k)$ to obtain good estimates of the source signals in $\mathbf{s}(k)$.

The definition of a "good estimate" in source separation, and the difficulty of the source-separation task in general, depend in large part on the forms of and knowledge about $\mathbf{s}(k)$, $\mathbf{A}$, and $\boldsymbol{\nu}(k)$. In this chapter, we restrict our discussion to problems in which the following assumptions hold:

- Each $s_i(k)$ in $\mathbf{s}(k)$ is statistically independent of $s_j(k)$ for $i \neq j$ and of $s_p(l)$ for any $0 \leq p \leq m-1$ and $l \neq k$. In addition, the probability density function (pdf) of $s_i(k)$ is the same as that of $s_i(l)$ for all k and l.
- The rank of $\mathbf{A}$ is greater than or equal to m and $m \leq n$.
- Each $v_i(k)$ is identically distributed, jointly Gaussian, and uncorrelated with $v_j(k)$ for $i \neq j$ and of $v_p(l)$ for any $0 \leq p \leq m-1$ and $l \neq k$. In addition, the elements of $\mathbf{v}(k)$ are independent of those of $\mathbf{s}(l)$ for all k and l.

These assumptions collectively guarantee that each measurement vector $\mathbf{x}(k)$ relays an equivalent level of *a priori* information regarding the source-signal samples in $\mathbf{s}(k)$. The second assumption guarantees that no information about the sources in $\mathbf{s}(k)$ is lost via the channel matrix $\mathbf{A}$. The third assumption guarantees the optimality of quadratic error criteria when training data are available, and appears in other formulations to this problem when discrete-amplitude source signals are present, such as in Grellier and Comon (1998).

3.2.2 Deconvolution

We now describe the task of deconvolution (Haykin 1994). Figure 3.1*b* shows the structure of this task, in which a discrete-time received signal $x(k)$ is assumed to be generated from an unknown source sequence $s(k)$ as

$$x(k) = \sum_{j=-\infty}^{\infty} a_j s(k-j) + v(k) \tag{3.2}$$

where $\{a_j\}$, $-\infty < j < \infty$ is the impulse response of the unknown linear time-invariant channel $A(z)$, defined as

$$A(z) = \sum_{j=-\infty}^{\infty} a_j z^{-j} \tag{3.3}$$

and $v(k)$ is a noise sequence. In this case, the goal is to process the sequence $x(k)$ to obtain good estimates of the source sequence $s(k)$.

As in the source-separation problem, we enforce certain assumptions and *a priori* knowledge of $s(k), A(z)$, and $v(k)$, as given below:

- Each $s(k)$ is statistically independent of $s(l)$ for $l \neq k$. In addition, the pdf of $s(k)$ is the same as that of $s(l)$ for all k and l.
- $A(z)$ is invertible; that is, there exists a sequence w_j such that

$$\sum_{j=-\infty}^{\infty} w_j a_{k-j} = c\delta(k - \Delta) \tag{3.4}$$

for some nonzero constant c and integer Δ.
- $A(z)$ is bounded-input, bounded-output stable; that is,

$$\sum_{j=-\infty}^{\infty} |a_j| < \infty \tag{3.5}$$

- Elements of the sequence $v(k)$ are uncorrelated Gaussian with zero mean and variance σ_v^2. In addition, $v(k)$ is independent of $s(l)$ for all k and l.

As in source separation, the preceding assumptions imply an equivalence of *a priori* information about the sequence $s(k)$ provided by each sample in $x(k)$. The second and third assumptions guarantee the information preserving and stability of $A(z)$, and the fourth assumption is common in formulations of this task (Haykin 1994).

3.2.3 Additional Information

In addition to these assumptions, additional information regarding the source-separation or deconvolution task may be available that make practical solutions to each task more easily implemented. Examples of this information in the source-separation task include:

1. Each $s_i(k)$ has the same pdf;
2. One or more of the $s_i(k)$ takes on discrete values in a finite alphabet $\mathcal{S}$;
3. One or more of the pdf's of the elements of $\mathbf{s}(k)$ are known;
4. The rank of $\mathbf{A}$ is known;
5. The column space of $\mathbf{A}$ is known;
6. The correlation matrix $\mathbf{R}_{xx} = E\{\mathbf{x}(k)\mathbf{x}^{\dagger}(k)\}$ is known; and
7. The number of sources m is known.

Examples of this information in the deconvolution task include:

1. $s(k)$ takes on discrete values in a finite alphabet $\mathcal{S}$;
2. The pdf of $s(k)$ is known;
3. The approximate nonzero support of the channel impulse response a_l is known; and
4. The magnitude response of the channel is known or, equivalently, $r_{xx}(l) = E\{x(k)x^*(k-l)\}$ is known.

3.2.4 Comparison

Considering the preceding formulations, we note two important distinctions regarding the nature of source separation and deconvolution:

- *The dimension of the source-separation task is at most n, whereas the dimension of the deconvolution task is generally greater than the length of the nonzero support of the channel impulse response a_j.* In other words, solutions to the source-separation task only require the processing of the elements of $\mathbf{x}(k)$ to obtain estimates of the elements in $\mathbf{s}(k)$. In the deconvolution task, optimally extracting $s(k)$ at each time instant usually requires the processing of at least as many samples of $x(k)$ in which $s(k)$ appears and typically involve many more samples to be processed if the channel $A(z)$ is highly frequency-selective. This fact implies that practical solutions to the deconvolution task often involve some amount of approximation in order to keep the complexity of the processing down to a reasonable level. In this respect, deconvolution is more difficult to solve than is source separation.
- *Source separation can involve m different source pdf's, whereas deconvolution only involves one source pdf.*[2] The potential difference between pdf's in source separation imply that solutions to the problem may require the estimation of different source pdf's, or the detection of differences between the source pdf's, in order to function properly. In contrast, one pdf completely characterizes the amplitude characteristics of $s(k)$ in the deconvolution task. In this respect, source separation is more difficult to solve than is deconvolution.

[2] In actuality, one could formulate source-separation and deconvolution tasks involving *time-varying* pdf's, although this extension is typically not considered in practice.

Historically, it is the latter issue—the multiplicity of pdf's in source separation—that perhaps has made BSS more difficult to solve than blind deconvolution.

3.3 ALGORITHMIC RELATIONSHIPS

3.3.1 Choice of Estimation Model

Obtaining estimates of the source signals in each of the deconvolution and source-separation tasks requires the choice of a computational model for processing the received signals. Although many structures are possible, the ones most often used are linear in form and are given by

$$\mathbf{y}(k) = \mathbf{W}\mathbf{x}(k) \tag{3.6}$$

and

$$y(k) = \sum_{j=-\infty}^{\infty} w_j x(k-j) \tag{3.7}$$

respectively, where the $(m \times n)$-dimensional matrix $\mathbf{W}$ and infinite-element sequence w_j contain the parameters of the source-separation and deconvolution models, respectively. We shall refer to the matrix $\mathbf{W}$ and the sequence w_j as the *separating matrix* and *deconvolution filter*, respectively. Since Eq. (3.7) represents discrete-time convolution, we can also express this relationship using z-transform notation as

$$Y(z) = W(z)X(z) \tag{3.8}$$

where $X(z)$, $Y(z)$, and $W(z)$ are the z-transforms of $x(k)$, $y(k)$, and w_k, respectively. Block diagrams of the source-separation and deconvolution systems are shown in Fig. 3.1*a* and *b*, respectively.

It is straightforward to show that both models in Eqs. (3.6) and (3.7) are sufficient to extract $\mathbf{s}(k)$ and $s(k)$ in the absence of noise ($\sigma_\nu^2 = 0$). These models are also reasonable when Gaussian-distributed interferences are present due to the optimality of linear estimators as conditional-mean estimators in such cases (Haykin 1996).

3.3.2 Choice of Cost Function

Given a parametric model for estimating the source signal(s) in $\mathbf{s}(k)$ or $s(k)$, one must choose a cost function with which to optimize the parameter values in $\mathbf{W}$ or $W(z)$. We review cost functions employing training data before exploring unsupervised error criteria.

Trained Criteria Although the subject of interest in this text is unsupervised adaptation methods, it is instructive to consider the solutions for $\mathbf{W}$ or $W(z)$ when all aspects of the problem are known, including source distributions, channel parameters, and noise characteristics. In such cases, two common criteria for which $\mathbf{W}$ or $W(z)$ can be optimized are *zero-forcing* (*ZF*) and *minimum mean-squared error* (*MMSE*). As an aside, both criteria produce identical solutions when observation noise is absent ($\sigma_v^2 = 0$).

The ZF solution for $\mathbf{W}$ or $W(z)$ causes each element of $\mathbf{y}(k)$ or $\{y(k)\}$ to contain only one element of $\mathbf{s}(k)$ or $\{s(k)\}$, respectively, such that the signal-to-interference ratio of each extracted source is maximized. Mathematically, we express this solution as

$$\mathbf{W}_{\text{ZF}} = \mathbf{PD}[\mathbf{A}^{\dagger}\mathbf{A}]^{-1}\mathbf{A}^{T} \tag{3.9}$$

for the source separation case, and

$$W_{\text{ZF}}(z) = cH^{-1}(z)z^{-\Delta} \tag{3.10}$$

for the deconvolution case, where $\mathbf{P}$ and $\mathbf{D}$ are m-dimensional permutation and nonsingular diagonal scaling matrices, respectively; c is a nonzero constant; Δ is an integer; and the superscript $\dagger$ denotes Hermitian transposition. If such a solution is obtained, then $\mathbf{y}(k)$ and $y(k)$ in the absence of noise are of the form

$$\mathbf{y}(k) = \mathbf{PDs}(k) \tag{3.11}$$

and

$$y(k) = cs(k - \Delta) \tag{3.12}$$

respectively.

By contrast, the MMSE solution for $\mathbf{W}$ or $W(z)$ maximizes the signal-to-noise ratio (SNR) of each of the elements of $\mathbf{s}(k)$ or $\{s(k)\}$ in $\mathbf{y}(k)$ or $\{y(k)\}$, respectively. These solutions minimize the cost functions

$$\mathcal{J}_{\text{MSE}}(\mathbf{W}) = E\{\|\mathbf{PDs}(k) - \mathbf{y}(k)\|^2\} \tag{3.13}$$

and

$$\mathcal{J}_{\mathrm{MSE}}(W(z)) = E\{|cs(k-\Delta) - y(k)|^2\} \tag{3.14}$$

respectively. The solutions for $\mathbf{W}$ and $W(z)$ are the well-known Wiener solutions for these respective tasks (Haykin 1996). In the source-separation case, this solution has the form

$$\mathbf{W}_{\mathrm{MSE}} = \mathbf{PDR_{sx}R_{xx}^{+}}, \tag{3.15}$$

where $\mathbf{R_{sx}}$ and $\mathbf{R_{xx}}$ are given by

$$\mathbf{R_{sx}} = E\{\mathbf{s}(k)\mathbf{x}^{\dagger}(k)\} = \mathbf{R_{ss}A}^{\dagger} \tag{3.16}$$

and

$$\mathbf{R_{xx}} = E\{\mathbf{x}(k)\mathbf{x}^{\dagger}(k)\} = \mathbf{AR_{ss}A}^{\dagger} + \sigma_v^2\mathbf{I} \tag{3.17}$$

respectively; $\mathbf{R_{ss}}$ is a diagonal matrix; and the superscript + denotes the matrix pseudoinverse. In the deconvolution case, the solution in the frequency domain is given by the spectral factorization of the cross-spectral density of $s(k)$ and $x(k)$ divided by the power spectrum of $x(k)$; see Haykin (1996) for details.

Note that these solutions indicate that the scales of the signals in $\mathbf{s}(k)$ and $s(k)$ are theoretically immaterial to the solutions for each task. In addition, the order of the sources in $\mathbf{s}(k)$ and the delay between $s(k)$ and $y(k)$ is usually not important, although one can obtain this information by analyzing the content of $\mathbf{y}(k)$ and $y(k)$ if the content of $\mathbf{s}(k)$ and $s(k)$ is identifiable.

In practice, complete knowledge of the channel and sources is usually not available; rather, K pairs of training signals given by $\{\mathbf{s}(k), \mathbf{x}(k)\}$ or $\{s(k), x(k)\}$ for $1 \leq k \leq K$ are all that we have. In such cases, these signals can be used to estimate the quantities needed to form the ZF and MMSE solutions. For example, a least-squares sample estimate of the channel $\mathbf{A}$ in the source-separation case could be calculated as

$$\hat{\mathbf{A}}_{LS} = \hat{\mathbf{R}}_{\mathbf{sx}}^{\dagger}\hat{\mathbf{R}}_{\mathbf{ss}}^{-1} \tag{3.18}$$

where

$$\hat{\mathbf{R}}_{\mathbf{sx}} = \frac{1}{K}\sum_{k=1}^{K} \mathbf{s}(k)\mathbf{x}^{\dagger}(k) \tag{3.19}$$

$$\hat{\mathbf{R}}_{\mathbf{ss}} = \frac{1}{K}\sum_{k=1}^{K} \mathbf{s}(k)\mathbf{s}^{\dagger}(k) \tag{3.20}$$

from which the value of $\mathbf{W}_{ZF}$ in Eq. (3.9) could be estimated. A least-squares estimate of the MMSE solution in Eq. (3.15) is given by

$$\hat{\mathbf{W}}_{LS} = \hat{\mathbf{R}}_{\mathbf{sx}}\hat{\mathbf{R}}_{\mathbf{xx}}^{-1} \tag{3.21}$$

where

$$\hat{\mathbf{R}}_{\mathbf{xx}} = \frac{1}{K}\sum_{k=1}^{K}\mathbf{x}(k)\mathbf{x}^{\dagger}(k) \tag{3.22}$$

Unsupervised Criteria In the absence of a desired response signal, the parameters in $\mathbf{W}$ or $W(z)$ can only be adapted using the measured signals in the vector $\mathbf{x}(k)$ or the sequence $\{x(k)\}$. Although many unsupervised adaptive approaches have been proposed for these tasks, they all have one goal in common: to make the elements of $\mathbf{y}(k)$ or $\{y(k)\}$ statistically independent. Statistical independence implies that the pdf of $\mathbf{y}(k)$ is of the form

$$f_{\mathbf{y}}(\mathbf{y}) = f_{\mathbf{y}}(y_0, y_1, \ldots, y_{m-1}) = \prod_{i=0}^{m-1} f_i(y_i) \tag{3.23}$$

where $f_i(y_i)$ is the marginal pdf of $y_i(k)$. Similarly, if the sequence $y(k)$ is independent from sample to sample, any m elements drawn from this sequence have a pdf of the form in Eq. (3.23) for a fixed $f_i(y) = f(y)$. When noise is not present in the measured signals, such a solution is equivalent to the ZF and MMSE solutions given earlier.

Unsupervised cost functions for adapting $\mathbf{W}$ or $W(z)$ to yield independent outputs are typically grouped into two categories:

- *Contrast functions* that attempt to measure the degree of independence of the system outputs; and
- *Density-based cost functions* that attempt to measure the distance of the sample pdf's of the system outputs from a set of model pdf's.

Both concepts can, in fact, be related using the highly principled maximum-likelihood method for estimating the parameters of an unknown statistical model when the pdf's of the source signal(s) satisfy certain smoothness properties. We now discuss the maximum-likelihood formulation to the source-separation and deconvolution tasks. Many of these ideas are drawn from Benveniste et al. (1980), Cardoso (1997), and Donoho (1981).

Let $\hat{f}_{\mathbf{v}}(\mathbf{v}, \theta)$ with parameter vector θ denote a parametrized model of the pdf $f_{\mathbf{v}}(\mathbf{v})$ of a random vector $\mathbf{v}$. In source separation, $\mathbf{v}$ represents the elements of the measured vector $\mathbf{x}(k)$ for any given k, and in deconvolution, $\mathbf{v}$ is a one-element vector that contains one sample from the measured sequence $x(k)$. A maximum-likelihood procedure attempts to determine the values in θ to maximize the log-likelihood function

$$\mathcal{J}_{\mathrm{ML}}(\theta) = \int_{\mathbf{v}} f_{\mathbf{v}}(\mathbf{v}) \log \hat{f}_{\mathbf{v}}(\mathbf{v}, \theta)\, d\mathbf{v} = E\{\log \hat{f}_{\mathbf{v}}(\mathbf{v}, \theta)\} \tag{3.24}$$

In this case, data-dependent versions of $\mathcal{J}_{\mathrm{ML}}(\theta)$ are obtained by sample averaging of the instantaneous value of $\log \hat{f}_{\mathbf{v}}(\mathbf{v}, \theta)$ for several realizations of $\mathbf{v}$.

Noting that $\hat{f}_{\mathbf{v}}(\mathbf{v}, \theta) = [\hat{f}_{\mathbf{v}}(\mathbf{v}, \theta)/f_{\mathbf{v}}(\mathbf{v})] f_{\mathbf{v}}(\mathbf{v})$, we have

$$\mathcal{J}_{\mathrm{ML}}(\theta) = -K(f_{\mathbf{v}} \| \hat{f}_{\mathbf{v}}) - H(f_{\mathbf{v}}) \tag{3.25}$$

where $H(f_{\mathbf{v}})$ is the entropy of $\mathbf{v}$ and

$$K(f_{\mathbf{v}} \| \hat{f}_{\mathbf{v}}) = \int_{\mathbf{v}} \log\left(\frac{f_{\mathbf{v}}(\mathbf{v})}{\hat{f}_{\mathbf{v}}(\mathbf{v}, \theta)}\right) f_{\mathbf{v}}(\mathbf{v})\, d\mathbf{v} \tag{3.26}$$

is the Kullback-Leibler divergence between the two pdf's $f_{\mathbf{v}}(\mathbf{v})$ and $\hat{f}_{\mathbf{v}}(\mathbf{v}, \theta)$ (Cardoso 1997). Since $H(f_{\mathbf{v}})$ does not depend on θ, maximum likelihood is equivalent to minimizing the Kullback-Leibler divergence between the unknown and model pdf's for a set of measurements. Note that such a minimization procedure is well posed even when $\hat{f}_{\mathbf{v}}(\mathbf{v}, \theta)$ does not equal $f_{\mathbf{v}}(\mathbf{v})$ for some value of θ, so long as $\hat{f}_{\mathbf{v}}(\mathbf{v}, \theta) > 0$ whenever $f_{\mathbf{v}}(\mathbf{v}) > 0$.

In maximum-likelihood formulations for solving the source-separation and deconvolution tasks, $\hat{f}_{\mathbf{v}}(\mathbf{v}, \theta)$ generally has two components:

1. Estimates of the linear channel $\mathbf{A}$ or $A(z)$; and
2. The marginal distributions of the signals in the vector $\mathbf{s}(k)$ or the sequence $\{s(k)\}$.

Assume that that both $\mathbf{W}$ and $W(z)$ are invertible and that observation noise is not present. Define the combined channels

$$\mathbf{C} = \mathbf{W}\mathbf{A} \quad \text{and} \quad C(z) = W(z)A(z) \tag{3.27}$$

respectively. Then, we can reparameterize the maximum-likelihood task in terms of the combined channel parameters rather than the unknown channel parameters, as the underlying structure of the estimation task is clearly unchanged in this case. With this reparameterization, $\mathbf{v}$ in Eq. (3.24) becomes $\mathbf{s}(k)$ or $s(k)$ for the source-separation and deconvolution tasks, respectively, and $\hat{f}_{\mathbf{v}}(\mathbf{v})$ becomes the observed sample pdf's of the signals at the output of the separation or deconvolution system, respectively. Thus, the maximum-likelihood approach attempts to adjust $\mathbf{C}$ or $C(z)$ to restore at the output of the system a distribution that is as close as possible to the source distribution. As an aside, such a technique only makes sense when the signals in $\mathbf{s}(k)$ or $\{s(k)\}$ are non-Gaussian, as $\mathbf{y}(k)$ or $y(k)$ are guaranteed to be Gaussian for any value of $\mathbf{C}$ or $C(z)$ for Gaussian-distributed sources (Benveniste et al. 1980; Cardoso 1997; Donoho 1981).

With regard to the choice of the output signal pdf model, we have two possible scenarios:

1. *The adjustable parameters only correspond to the combined channel or mixture.* Then, the hypothesis regarding the marginal distribution(s) of the source(s) is fixed; that is, the computational form of $\hat{f}_{\mathbf{v}}(\cdot)$ is not adjusted within the parameter-estimation procedure. Such a choice is reasonable when the source distributions are known *a priori*. Alternatively, if an approximate form of $f_{\mathbf{v}}(\cdot)$ is used, then the parameter estimates obtained in a maximum-likelihood procedure can be consistent, but are not statistically efficient in general. The latter case is typical of contrast-based approaches, in which a simple function such as the signal kurtosis is used in place of a likelihood-based measure (Comon 1994; Delfosse and Loubaton 1995).
2. *The adjustable parameters include both the combined channel the source distribution.* If a parametric model is chosen for the form of $\hat{f}_{\mathbf{v}}(\cdot)$, then maximum likelihood allows for the optimization of all of the parameters in a consistent and specified fashion. An alternative approach is to employ a parametric family of estimating functions whose averaged zero points provide consistent estimates of the parameters in $\mathbf{W}$ or $W(z)$, in which the parameters of the estimating functions are determined from the measurements. A general discussion of estimating functions as applied to the BSS task can be found in Amari and Cardoso (1997), and an example of this approach is described in Pham et al. (1992).

3.3.3 Choice of Adaptive Algorithm

Once a cost function is chosen for the separation or deconvolution system, an algorithm for adjusting the parameters in $\mathbf{W}$ or $W(z)$ must be chosen. Typically, an unconstrained or constrained stochastic-gradient technique is employed, as these methods are often simpler to implement than more sophisticated Gauss-Newton or conjugate-gradient optimization techniques that make use of second-order information about the cost-function error surface. As an example, an unconstrained stochastic-gradient approach for the source-separation task adjusts the (i, j)th element of the separation system as

$$w_{ij}(k+1) = w_{ij}(k) - \mu(k) \frac{\partial \hat{\mathcal{J}}(\mathbf{W}(k))}{\partial w_{ij}} \tag{3.28}$$

where $\hat{\mathcal{J}}(\mathbf{W}(k))$ is an instantaneous estimate of some averaged cost function $E\{\hat{\mathcal{J}}(\mathbf{W})\}$ evaluated at $\mathbf{W} = \mathbf{W}(k)$, and $\mu(k)$ is a small positive step-size sequence. The forms of constrained gradient search methods depend on the type of constraint imposed on the parameter space. For a survey of unit-norm constrained gradient methods that are useful for contrast-based approaches to source separation and deconvolution, see Douglas et al. (1998).

Typically, the value of the instantaneous cost function appearing within the gradient updates depends on some nonlinear function of the output signals in $\mathbf{y}(k)$ or $y(k)$. If maximum likelihood is used, then the score function(s)

$$\phi_i(y) = -\frac{\partial \log f_i(y)}{\partial y} \tag{3.29}$$

appear within the coefficient updates in a gradient-based method. If a parametric model of the source-signal distribution(s) is (are) chosen, then a gradient-based procedure updates both the separator or equalizer coefficients as well as the an estimate $\hat{\phi}_i(y)$ of each $\phi_i(y)$ as calculated using samples from the output sequences $y(k)$ or $\mathbf{y}(k)$. Although accurate estimates of $\hat{\phi}_i(y)$ are desired for statistical efficiency, it should be noted that successful separation or deconvolution can be obtained even when $\hat{\phi}_i(y)$ differs from $\phi_i(y)$ in Eq. (3.29). If the deviation is too large, however, then a gradient algorithm can fail to separate or deconvolve the source signals. In general, the stability of a chosen minimization or maximization procedure about a separating solution depends on the form of $\phi_i(y)$ as well as the source-signal distribution of the ith extracted source. Details of these issues can be found in Amari et al. (1997a), Benveniste et al. (1980), and Cardoso and Laheld (1996). Examples in the next section discuss this issue in more detail in the context of one particular blind deconvolution algorithm (Amari et al. 1997b,c).

Once an algorithm for adjusting $\mathbf{W}$ or $\{w_i\}$ has been specified, convergence to a source-separation or deconvolution condition can be problematic due to the nonquadratic nature of the cost-function error surface. In particular, convergence is often slow for standard gradient-based optimization methods. Whitening the input-signal measurements prior to the application of the separator or equalizer can improve convergence performance, as each system can then be restricted to that of a rotation in coefficient space (Benveniste et al. 1980; Comon 1994; Delfosse and Loubaton 1995; Kung and Mejuto 1998). The recently developed relative or natural-gradient search methods have shown excellent convergence performance in these cases, particularly for ill-conditioned channels. A discussion of these methods as applied to BSS and blind deconvolution can be found in Amari et al. (1996), Cardoso and Laheld (1996), and Cichocki et al. (1994) and Amari et al. (1997b,c) and Douglas et al. (1996), respectively, and one such algorithm is discussed in the next section.

3.4 STRUCTURAL RELATIONSHIPS

We now consider the structural relationship between the blind deconvolution and BSS tasks. To draw a link between these two problems and useful algorithms for their solution, we formulate a new BSS task in which both $\mathbf{A}$ and $\mathbf{W}(k)$ exhibit a circulant structure. Note that solutions to source separation under circulant mixing conditions should not be seen as ends in and of themselves; rather, they are useful in that they can be quite easily translated to algorithms for blind deconvolution tasks. As an example, we consider density-based BSS algorithms for extracting signals under circulant mixing conditions. Using the derivation method in Amari et al. (1996) and Cardoso and Laheld (1996), we provide both stochastic and natural/relative gradient density-based algorithms for source separation in this case, and we relate them to the results in Lambert (1996). Then, via, limiting arguments, we extend the algorithms to the infinite-dimensional case, where circulant matrices become filtering (Toeplitz) matrices, in which single-channel blind deconvolution algorithms are obtained. These algorithms are identical to those described in Amari et al. (1997b,c), although our derivation method uses somewhat simpler tools than those used in Amari et al. (1997b,c). Extensions of the methods to contrast-based and multidimensional blind deconvolution tasks are also mentioned.

3.4.1 Source Separation for Circulant Mixing Conditions

Consider the source-separation task in Fig. 3.1a with $m = n$, except that the matrix $\mathbf{A}$ exhibits a circulant structure given by

$$
\mathbf{A} = \begin{bmatrix}
a_0 & \cdots & a_{-M} & 0 & \cdots & 0 & a_M & \cdots & a_1 \\
\vdots & \ddots & & \ddots & & & & \ddots & \vdots \\
 & & \ddots & & \ddots & & & & a_M \\
a_M & & a_0 & & a_{-M} & & & & 0 \\
0 & \ddots & & \ddots & & & \ddots & & \vdots \\
\vdots & & \ddots & & \ddots & & & \ddots & 0 \\
0 & & a_M & & a_0 & & & & a_{-M} \\
a_{-M} & & & \ddots & & & \ddots & & \\
\vdots & \ddots & & & \ddots & & & & \vdots \\
a_{-1} & \cdots & a_{-M} & 0 & \cdots & 0 & a_M & \cdots & a_0
\end{bmatrix} \tag{3.30}
$$

Note that such a situation is not typically observed in practice, although the algorithms developed for this task can be easily translated to blind deconvolution problems, as will be shown. Since $\mathbf{A}$ is a circulant matrix, then $x_i(k)$ is

$$
x_i(k) = \sum_{p=-M}^{M} a_p s_{[i-p]_n}(k) \tag{3.31}
$$

where $[\cdot]_n$ denotes the modulo-n operation. Thus, $\mathbf{x}(k)$ is obtained from the circular convolution of the channel impulse response $a_j, -M \leq j \leq M$, and the source sequence in $\mathbf{s}(k)$. To extract the source sequence, we apply an $(n \times n)$ circulant demixing matrix $\mathbf{W}(k)$ with entries given by

$$
[\mathbf{W}(k)]_{ij} = \begin{cases} w_{i-j}(k) & \text{if } [|i-j|]_n \leq L \\ 0 & \text{otherwise} \end{cases} \tag{3.32}
$$

to $\mathbf{x}(k)$ to produce $\mathbf{y}(k)$, as in Eq. (3.6).

The only difference between this and the original source-separation task is the circulant structure of **A** and $\mathbf{W}(k)$. Note that the product of two circulant matrices is also a circulant matrix. The goal of this task is to adjust $\mathbf{W}(k)$ such that

$$\mathbf{W}(k)\mathbf{A} \approx c\mathbf{I}_\Delta \tag{3.33}$$

where c is a nonzero constant, and $\mathbf{I}_\Delta$ is a circulant matrix whose Δth lower diagonal has all unity elements. The unknown quantities c and Δ represent the fact that we can only extract the source sequence up to an arbitrary scaling and circulant delay, although it is the case that the source samples will be extracted in circulant order because of the problem's circulant structure. The form of $\mathbf{W}(k)$ is adequate to separate the source sequence; that is, there always exists a sequence $\{w_j(k)\}$ and a value of $L, 2L+1 \leq n$, such that Eq. (3.33) can be obtained.

3.4.2 Density-Based Methods for Circulant Mixing Conditions

We first consider density-based methods for instantaneous BSS as applied to the system in Eq. (3.31). To adjust $\mathbf{W}(k)$, we employ the approximate negative log-likelihood function given by Amari et al. (1996), Bell and Sejnowski (1995), and Cardoso (1997).

$$\hat{\mathcal{J}}(\mathbf{W}) = -\log\left[|\det \mathbf{W}| \prod_{i=1}^{n} f(y_i)\right] \tag{3.34}$$

$$= -\log|\det \mathbf{W}| - \sum_{i=1}^{n} \log\hat{f}(y_i) \tag{3.35}$$

and $\hat{f}(y)$ is an estimate of the pdf of the source signal. The stochastic-gradient algorithm for minimizing $E\{\hat{\mathcal{J}}(\mathbf{W})\}$ is

$$w_j(k+1) = w_j(k) - \frac{\mu(k)}{n}\frac{\partial\hat{\mathcal{J}}(\mathbf{W}(k))}{\partial w_j} \tag{3.36}$$

for $-L \leq j \leq L$, where $\mu(k)$ is a chosen step size. It can be shown that

$$\frac{\partial}{\partial w_j}\sum_{i=0}^{n-1} \log\hat{f}(y_i) = -\sum_{i=0}^{n-1} \hat{\phi}(y_i)x^*_{[i-j]_n} = -\mathrm{tr}_j[\hat{\phi}(\mathbf{y})\mathbf{x}^\dagger] \tag{3.37}$$

where $\hat{\phi}(\mathbf{y}) = [\phi(y_1)\cdots\phi(y_n)]^T$, $\phi(y) = -\partial\log\hat{f}(y)/\partial y$, and $\mathrm{tr}_j[\mathbf{M}]$ is the sum of the jth lower circulant diagonal of the matrix $\mathbf{M}$.

To determine the derivative of $\log|\det \mathbf{W}|$, we note that

$$\mathbf{W} = \sum_{i=-L}^{L} w_i \mathbf{I}_i \tag{3.38}$$

For any nonsingular matrix $\mathbf{A}$, we have $d \log|\det \mathbf{A}| = \mathrm{tr}_0[\mathbf{A}^{-*} d\mathbf{A}]$, such that

$$d \log|\det \mathbf{W}| = \sum_{j=-L}^{L} dw_j \, \mathrm{tr}_{-j}[\mathbf{W}^{-*}] \tag{3.39}$$

Therefore,

$$\frac{\partial \log|\det \mathbf{W}|}{\partial w_j} = \mathrm{tr}_j[\mathbf{W}^{-\dagger}] \tag{3.40}$$

Thus, the update for each unique entry of $\mathbf{W}(k)$ is

$$w_j(k+1) = w_j(k) + \frac{\mu(k)}{n} \mathrm{tr}_j[\mathbf{W}^{-\dagger}(k) - \hat{\phi}(\mathbf{y}(k))\mathbf{x}^{\dagger}(k)] \tag{3.41}$$

Since $w_j(k) = \mathrm{tr}_j[\mathbf{W}(k)]/n$, we could equivalently write

$$\mathrm{tr}_j[\mathbf{W}(k+1)] = \mathrm{tr}_j[\mathbf{W}(k) + \mu(k)\{\mathbf{W}^{-\dagger}(k) - \hat{\phi}(\mathbf{y}(k))\mathbf{x}^{\dagger}(k)\}] \tag{3.42}$$

Equation (3.42) closely resembles the equivalent stochastic-gradient algorithm for standard BSS (Amari et al. 1996; Bell and Sejnowski 1995).

As in the original source-separation task, the performance of Eq. (3.41) will be poor when $\mathbf{A}$ is ill-conditioned. We therefore seek an algorithm that has better convergence properties. From developments in source separation (Amari et al. 1996; Cardoso and Laheld 1996), it is reasonable to consider the modified algorithm

$$w_j(k+1) = w_j(k) + \frac{\mu(k)}{n} \mathrm{tr}_j[\{\mathbf{W}^{-\dagger}(k) - \hat{\phi}(\mathbf{y}(k))\mathbf{x}^{\dagger}(k)\}\mathbf{W}^{\dagger}(k)\mathbf{W}(k)] \tag{3.43}$$

in analogy with the relative- and natural-gradient algorithms. The resulting update for each $w_j(k)$ is

$$w_j(k+1) = w_j(k) + \mu(k)\left\{ w_j(k) - \frac{1}{n}\sum_{i=0}^{n-1} \hat{\phi}(y_i(k)) u^*_{[i-j]_n}(k) \right\} \tag{3.44}$$

where we have defined

$$\mathbf{u}(k) = [u_0(k)u_1(k)\cdots u_{n-1}(k)]^T = \mathbf{W}^\dagger(k)\mathbf{y}(k) \qquad (3.45)$$

Is such a modification of the algorithm justified? Consider the following points:

1. Circulant matrices whose eigenvalues are nonzero and whose entries are otherwise unconstrained are closed under addition and multiplication; hence, they form a group. In standard source separation, the group structure of the parameter space can be exploited to derive the natural- or relative-gradient algorithms for these problems (Amari et al. 1996; Cardoso and Laheld 1996).
2. With this modification, the update can be written as

$$\text{tr}_j[\mathbf{W}(k+1)] = \text{tr}_j[\mathbf{W}(k) + \mu(k)\{\mathbf{I}_0 - \hat{\phi}(\mathbf{y}(k))\mathbf{y}^\dagger(k)\}\mathbf{W}(k)] \quad (3.46)$$

It is possible to prove for two matrices $\mathbf{A}$ and $\mathbf{B}$ and a third circulant matrix $\mathbf{C}$ that if $\text{tr}_j[\mathbf{A}] = \text{tr}_j[\mathbf{B}]$, then $\text{tr}_j[\mathbf{AC}] = \text{tr}_j[\mathbf{BC}]$. Using Eq. (3.27), Eq. (3.46) is

$$\text{tr}_j[\mathbf{C}(k+1)] = \text{tr}_j[\mathbf{C}(k) + \mu(k)\{\mathbf{I}_0 - \hat{\phi}(\mathbf{y}(k))\mathbf{y}^\dagger(k)\}\mathbf{C}(k)] \quad (3.47)$$

$$\mathbf{y}(k) = \mathbf{C}(k)\mathbf{s}(k) \qquad (3.48)$$

which does not depend on $\mathbf{A}$. Hence, Eq. (3.44) is *equivariant* with respect to all circulant mixing matrices, a situation analogous to the standard source separation task (Cardoso and Laheld 1996).

These points suggest that the modification in Eq. (3.43) is appropriate when both $\mathbf{W}(k)$ and $\mathbf{A}$ are circulant.

3.4.3 Relationships to Blind Deconvolution Algorithms

We now relate the stochastic-gradient and equivariant algorithms in Eqs. (3.41) and (3.44) to existing algorithms for the blind deconvolution task. In our discussion, we assume that the elements of $\mathbf{s}(k)$ form a time series, such that

$$\mathbf{s}(k) = [s(k)s(k+1)\cdots s(k+n-1)]^T \qquad (3.49)$$

We first show Eqs. (3.41) and (3.44) as single-channel time-domain versions of the multichannel frequency-domain deconvolution algorithms described in Lambert (1996). To see this fact, note that the set of discrete Fourier transform (DFT) vectors

$$\mathbf{e}_i = [1 e^{-j2\pi i/n} \cdots e^{-j2\pi(n-1)i/n}]^T \tag{3.50}$$

for $0 \le i \le (n-1)$ form an orthonormal basis for any circulant matrix (Jain 1998). Assume that $n = 2L + 1 = 2M + 1$, and define the n-dimensional vectors $\mathscr{H}$ and $\mathscr{W}(k)$ as the DFTs of the sequences $\{a_0, \ldots, a_M, a_{-M}, \ldots, a_{-1}\}$ and $\{w_0(k), \ldots, w_L(k), w_{-L}(k), \ldots, w_{-1}\}$, respectively. Then, Eqs. (3.41) and (3.44) can be expressed as

$$\mathscr{W}(k+1) = \mathscr{W}(k) + \mu(k)\left\{\mathbf{1} \oslash \mathscr{W}^*(k) - \frac{1}{n}\hat{\Phi}(k) \odot \mathscr{X}^*(k)\right\} \tag{3.51}$$

and

$$\mathscr{W}(k+1) = \mathscr{W}(k) + \mu(k)\left\{\mathscr{W}(k) - \frac{1}{n}\hat{\Phi}(k) \odot \mathscr{U}^*(k)\right\} \tag{3.52}$$

$$\mathscr{U}(k) = \mathscr{W}^*(k) \odot \mathscr{Y}(k) = \mathscr{W}^* \odot \mathscr{W}(k) \odot \mathscr{X}(k) \tag{3.53}$$

respectively, where $\mathbf{1}$ is a vector of ones; $\odot$ and $\oslash$ denote point-by-point multiplication and division of two vectors; and $\mathscr{X}(k)$, $\mathscr{Y}(k)$, and $\hat{\Phi}(k)$ are the DFTs of the vectors $\mathbf{x}(k)$, $\mathbf{y}(k)$, and $\hat{\boldsymbol{\phi}}(\mathbf{y}(k))$, respectively. These expressions are of the same form as those in Lambert (1996). Our derivations indicate that such algorithms are only technically appropriate for circulant mixing conditions, however, and they must be modified for use in blind deconvolution tasks.

To translate Eq. (3.44) into an equivariant algorithm for blind deconvolution, we invoke a theory that relates the asymptotic forms of circulant and Toeplitz matrices as $n \to \infty$ (Gray 1977). Within such a framework, it can be shown that products of circulant matrices are similar to products of Toeplitz matrices, so long as the sequences defining the matrices are absolutely summable. Thus, circular convolutions become similar to linear convolutions in the limit, such that the elements of the vectors $\mathbf{x}(k)$, $\mathbf{y}(k)$, and $\mathbf{u}(k)$ become

$$x_i(k) = \sum_{j=-\infty}^{\infty} a_j s(k+i-j) \tag{3.54}$$

$$y_i(k) = \sum_{j=-\infty}^{\infty} w_j(k) x_{i-j}(k) \tag{3.55}$$

$$u_i(k) = \sum_{j=-\infty}^{\infty} w_j(k) y_{i+j}(k) \tag{3.56}$$

respectively. With these definitions, the coefficient updates in Eq. (3.44) translate to

$$w_j(k+1) = w_j(k) + \mu(k)\left[w_j(k) - \lim_{n\to\infty}\frac{1}{n}\sum_{i=0}^{n-1}\hat{\phi}(y_i(k))u^*_{i-j}(k)\right] \quad (3.57)$$

By the law of large numbers, the second term within brackets on the right-hand side (RHS) of Eq. (3.57) approaches $E\{\hat{\phi}(y_i(k))u_{i-j}(k)\}$. To obtain an instantaneous update, we replace the expected quantity with its instantaneous value for $i = k$, which yields

$$w_j(k+1) = w_j(k) + \mu(k)[w_j(k) - \hat{\phi}(y_k(k))u^*_{k-j}(k)] \quad (3.58)$$

The coefficient update in Eq. (3.58) is identical to the natural-gradient blind deconvolution algorithm in Amari et al. (1997b,c) in the single-channel case if differences in the coefficient delays within the update terms are ignored.

In practice, a truncated finite-duration impulse-response (FIR) filter is employed in place of the infinite-duration impulse-response (IIR) filters in Eqs. (3.55)–(3.56), and delayed signal values are used to form the coefficient update terms to maintain the causality of the overall system. Details of these approximations are given in Amari et al. (1997b) and Douglas et al. (1996, 1999b). The simplest algorithm so obtained is

$$y(k) = \sum_{l=0}^{L} w_l(k)x(k-l) \quad (3.59)$$

$$u(k) = \sum_{q=0}^{L} w^*_{L-q}(k)y(k-q) \quad (3.60)$$

$$w_j(k+1) = w_j(k) + \mu(k)\{w_j(k) - \hat{\phi}(y(k-L))u^*(k-j)\} \quad (3.61)$$

With these approximations, the equivariant performance of the algorithm is technically lost in general, and as in all blind deconvolution schemes, its effectiveness depends on how well the resulting filter can approximate a delayed inverse of the channel.

3.4.4 Stability Issues

The choice of the nonlinearity $\hat{\phi}(y)$ in Eq. (3.61) depends on the pdf $f_s(s)$ of each element of the source-signal sequence $s(k)$. It is possible to analyze the local stability characteristics of the density-based algorithm in Eq. (3.44) under circulant mixing conditions via the form of the Hessian of the cost function $E\{\mathcal{J}(\mathbf{W})\}$ about a separating solution satisfying Eq. (3.33). For real-valued signals and coefficients, this situation is simply a special case of a more general algorithm form discussed in Cardoso and Laheld (1996) and Douglas et al. (1997). As such, we can translate the results in Cardoso and Laheld (1996) and Douglas et al. (1997) directly. For small values of $\mu(k)$, the algorithm adjusts $\mathbf{W}(k)$ such that

$$E\{y(k)\hat{\phi}(y(k))\} = 1 \tag{3.62}$$

at a separating solution. Moreover, this solution is locally stable if the following three conditions are satisfied:

$$E\{\hat{\phi}'(y(k))\} > 0 \tag{3.63}$$

$$E\{y^2(k)\hat{\phi}'(y(k))\} + E\{y(k)\hat{\phi}(y(k))\} > 0 \tag{3.64}$$

$$E\{y(k)\hat{\phi}(y(k))\} - E\{y^2(k)\}E\{\hat{\phi}'(y(k))\} < 0 \tag{3.65}$$

The first two of these conditions are normally satisfied in practice for any odd-symmetric nondecreasing function $\hat{\phi}(y)$. The third condition places a restriction on the types of sources that can be extracted for any given choice for $\hat{\phi}(y)$. In particular, the class of signals for which the algorithm performs deconvolution for a given $\hat{\phi}(y)$ is much larger than the source signal whose pdf satisfies $\hat{\phi}(y) = -\partial \log f_s(y)/\partial y$. The two following examples illustrate this issue.

Example 1. Suppose all quantities are real-valued and $\hat{\phi}(y) = y^3$ is chosen. This choice corresponds to the density model

$$f_s(s) = Ke^{\alpha|s|^4} \tag{3.66}$$

where K and α are distribution constants. Then, the condition in Eq. (3.65) corresponds to

$$\sigma_y^2\left[\frac{E\{s^4(k)\}}{\sigma_s^4} - 3\right] < 0 \tag{3.67}$$

where $\sigma_y^2 = E\{|y(k)|^2\}$ and $\sigma_s^2 = E\{|s(k)|^2\}$, respectively. The term within brackets on the left-hand side (LHS) of Eq. (3.67) is called the *normalized kurtosis* of the source signal and is scale-independent, that is, the absolute magnitude of the sequence $\{s(k)\}$ can be arbitrary. Thus, choosing $\hat{\phi}(y) = y^3$ allows Eqs. (3.59)–(3.61) to deconvolve any negative-kurtosis signal. Source pdf's with negative kurtoses include many types of signals used in digital communication tasks (Treichler et al. 1998).

Example 2. Suppose the logistic function $\hat{\phi}(y) = \tanh(\beta y)$ is selected, corresponding to the density model

$$f_s(s) = K \log \cosh(\beta s) \tag{3.68}$$

for some distribution constant K. Then, the condition in Eq. (3.65) evaluates to

$$E\{y(k)\tanh(\beta y(k))\} - \beta E\{y^2(k)\}E\{\text{sech}^2(\beta y(k))\} < 0 \tag{3.69}$$

Moreover, as β tends to zero, we have

$$\lim_{\beta \to 0} \hat{\phi}(y) = \text{sgn}(y) \tag{3.70}$$

and the source pdf $f_s(s)$ must satisfy the limiting condition

$$\sigma_y \left[\frac{E\{|s(k)|\}}{\sigma_s} - 2\sigma_s f_s(0) \right] < 0 \tag{3.71}$$

for stability. The term in brackets on the LHS of Eq. (3.71) is scale independent and is typically negative valued for impulsive sources having "heavy-tailed" distributions.

3.4.5 Simulations

We now explore the behavior of the blind deconvolution algorithm in Eqs. (3.59)–(3.61) via simulations. Our example is one drawn from digital communication systems incorporating 16-bit quadrature-amplitude-modulated (16-QAM) signal constellations. A complex-valued 16-QAM source has a probability mass function (pmf) given by

$$p_s(s) = \begin{cases} \frac{1}{16} & \text{if } s \in \mathcal{S} \\ 0 & \text{otherwise} \end{cases} \tag{3.72}$$

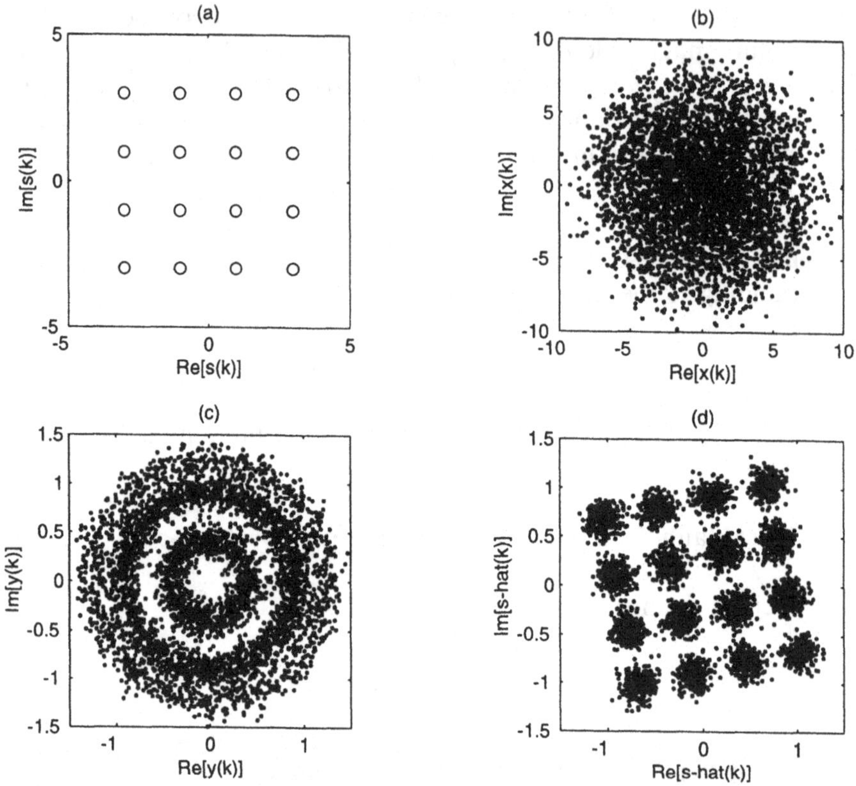

Figure 3.2 Signal constellations in the 16-QAM blind equalization simulation example: (*a*) source signal $s(k)$; (*b*) received signal $x(k)$; (*c*) equalizer output $y(k)$; and (*d*) carrier-offset-compensated output $\hat{s}(k)$.

where the 16-element set $\mathcal{S}$ contains all values of $s_R + js_I$ with $s_R \in \{-3, -1, 1, 3\}$ and $s_I \in \{-3, -1, 1, 3\}$. Figure 3.2*a* shows the elements of $\mathcal{S}$ as small circles on the complex plane. The measured signal $x(k)$ is modeled using a modified version of Eq. (3.2) as given by

$$x(k) = e^{j\delta k}\left[\sum_{j=-\infty}^{\infty} a_j s(k-j)\right] + v(k) \tag{3.73}$$

where a_j is given by

$$a_j = \begin{cases} 1.0 + j0.3 & \text{if } j = 0 \\ -0.5 + j1.0 & \text{if } j = 1 \\ 0.25 - j0.1 & \text{if } j = 2 \\ 0 & \text{otherwise} \end{cases} \tag{3.74}$$

and $v(k)$ is a zero-mean jointly Gaussian complex random sequence, and $\delta = 0.01$. The additional exponential factor on the RHS of Eq. (3.73) models the detrimental effects of a carrier offset (due to inexact knowledge of the carrier frequency) that are often present in QAM-based communication systems, and the variance of $v(k)$ was chosen to provide a 30-dB signal-to-noise ratio in the received signal. Figure 3.2*b* shows 5000 samples of such a complex baseband received signal $x(k)$, in which the amplitude structure of the source signal in Fig. 3.2*a* has clearly been lost.

To simulate the behavior of the algorithm, pseudorandom signals with the aforementioned statistical characteristics were created and processed via the algorithm using the MATLAB signal-analysis software package. In this case, $\hat{\phi}(y) = |y|^2 y$ was chosen for the equalizer nonlinearity, and the equalizer coefficients were initialized using a "center-spike" strategy in which

$$w_j = \delta\left(j - \frac{L}{2}\right) \tag{3.75}$$

In addition, a quasi-normalized step size of the form

$$\mu(k) = \frac{\mu_0}{\beta + \rho(k)} \tag{3.76}$$

$$\rho(k) = \lambda\rho(k-1) + (1-\lambda)|y(k)|^4 \tag{3.77}$$

was used in the coefficient updates. The choice of $\rho(k)$ is motivated by the form of the second term within brackets in the coefficient update of Eq. (3.61), which contains fourth-moment output terms of the form $\{|y(k-L)|^2 y(k-L) y^*(k-l)\}$ for our particular choice of $\hat{\phi}(y)$. The parameters chosen for the simulation were $L = 23$, $\mu_0 = 0.0008$, $\beta = 0.1$, $\lambda = 0.99$, and $\rho(0) = 1$.

Figure 3.3 shows the evolution of the intersymbol interference (ISI) computed as

$$\mathrm{ISI}(k) = \frac{\sum_{l=0}^{2L} |c_l(k)|^2}{\max_{0 \le j \le 2L} |c_j(k)|^2} - 1 \tag{3.78}$$

where $c_j(k)$ is the convolution of $w_j(k)$ and a_j. As can be seen, the algorithm converges to a low-ISI state in about 10,000 iterations. Shown in Fig. 3.2*c* are the last 5000 samples of $y(k)$ computed in the simulation.

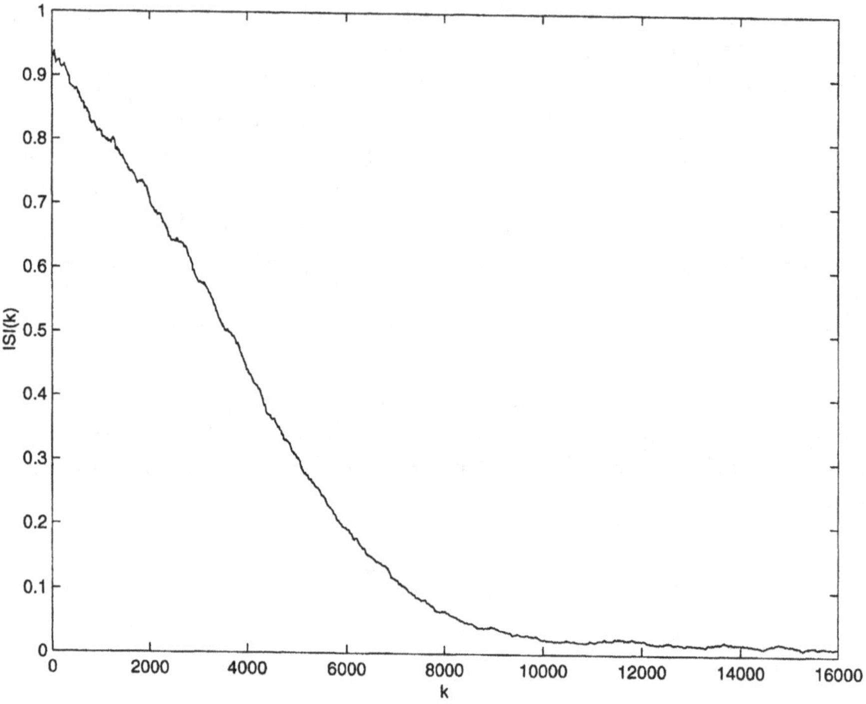

Figure 3.3 Evolution of the ISI in the 16-QAM blind equalization simulation example.

The concentric-ring structure in this plot is indicative of an equalized QAM constellation with a carrier offset present. Figure 3.2*d* shows the last 5000 samples of the "carrier-offset-compensated" signal

$$\hat{s}(k) = y(k)e^{-j\delta k} \tag{3.79}$$

in which a noisy, scaled, and rotated version of the original QAM signal pmf is observed. These results suggest that, with the proper choice of nonlinearity $\hat{\phi}(y)$, the algorithm is effective for blind deconvolution in digital communication systems, even when carrier offset and measurement noise are present.

3.5 EXTENSIONS

3.5.1 Contrast-Based Methods for Blind Deconvolution

A similar formulation as that used in Section 3.4 can be used to develop blind deconvolution methods from contrast-based approaches to source separation. Such an approach was used in Douglas and Kung (1998a) to extend a unit-norm-constrained kurtosis-based source-separation method to the blind deconvolution task. One important distinction between deconvolution and source separation in this case is that explicit deflation of the signal space is unnecessary, as once a particular estimate $y(k)$ of any $cs(k - \Delta)$ has been reliably extracted using a linear combiner model of the form

$$y(k) = \sum_{j=0}^{L} w_j x(k - j) \tag{3.80}$$

all the elements of the source sequence can be extracted by a linear filtering operation. From this viewpoint, all contrast-based approaches are immediate candidates for blind deconvolution algorithms, and only minor translations in notation are necessary.

3.5.2 Multichannel Extensions

The concepts described in Section 3.4 can be extended to the multiple-input, multiple-output (MIMO) case by defining a block-circulant matrix $\mathbf{A}$ of the form

$$\mathbf{A} = \begin{bmatrix} \mathbf{A}_0 & \cdots & \mathbf{A}_{-M} & 0 & \cdots & 0 & \mathbf{A}_M & \cdots & \mathbf{A}_1 \\ \vdots & \ddots & & \ddots & & & & \ddots & \vdots \\ & & \ddots & & \ddots & & & & \mathbf{A}_M \\ \mathbf{A}_M & & \mathbf{A}_0 & & & \mathbf{A}_{-M} & & & 0 \\ 0 & \ddots & & \ddots & & & \ddots & & \vdots \\ \vdots & & \ddots & & \ddots & & & \ddots & 0 \\ 0 & & \mathbf{A}_M & & & \mathbf{A}_0 & & & \mathbf{A}_{-M} \\ \mathbf{A}_{-M} & & & \ddots & & & \ddots & & \\ \vdots & \ddots & & & \ddots & & & & \vdots \\ \mathbf{A}_{-1} & \cdots & \mathbf{A}_{-M} & 0 & \cdots & 0 & \mathbf{A}_M & \cdots & \mathbf{A}_0 \end{bmatrix} \tag{3.81}$$

where each $\mathbf{A}_i$ is an $(n \times m)$-dimensional matrix. All of the resulting derivations extend to this case with minor modifications. For example, the multichannel versions of the density-based algorithm in Eqs. (3.59)–(3.61) derived in this manner is identical to that derived in Amari et al. (1997b) in the space of infinite-dimensional discrete-time systems.

From these manipulations, the following rules regarding the extension of $(m \times n)$-dimensional spatial-only adaptive algorithms to n-input, m-output multidimensional dispersive systems can be intuited, as described in Douglas et al. (1999a).

- Multiplication of two matrices in the spatial-only case is equivalent to convolution of their associated matrix sequences in the multichannel dispersive case.
- Addition of two matrices in the spatial-only case is equivalent to element-by-element addition of their associated matrix sequences in the multichannel dispersive case.
- Transposition of a matrix in the spatial-only case is equivalent to element-by-element transposition and time-reversal of its associated matrix sequence in the multichannel dispersive case.

These rules have been used to derive generic forms of stochastic-gradient adaptive paraunitary filters (Douglas et al. 1999a).

3.6 CONCLUSIONS

We have considered the algorithmic and structural similarities of the blind deconvolution and BSS tasks. Maximum likelihood is shown to provide a unified principle on which successful methods for both tasks can be based. Moreover, the structural relationships between the two tasks are illustrated by considering the problem of and solutions for the blind separation of circulant mixtures, in which recently developed algorithms for blind deconvolution are readily obtained. From this study, it is apparent that blind deconvolution and BSS are related problems, and by drawing from this correspondence, the possibility for the development of new and innovative solutions to both tasks is indicated.

REFERENCES

Amari, S., and J.-F. Cardoso, Nov. 1997, "Blind source separation—Semiparametric statistical approach," *IEEE Trans. Signal Processing*, vol. 45, pp. 2692–2700.

Amari, S., A. Cichocki, and H. H. Yang, 1996, "A new learning algorithm for blind signal separation," *Adv. Neural Inform. Proc. Sys. 8* (Cambridge, MA: MIT Press), pp. 757–763.

Amari, S., T.-P. Chen, and A. Cichocki, Nov. 1997a, "Stability analysis of learning algorithms for blind source separation," *Neural Networks*, vol. 10, pp. 1345–1351.

Amari, S., S. C. Douglas, A. Cichocki, and H. H. Yang, Apr. 1997b, "Multichannel blind deconvolution using the natural gradient," *Proc. IEEE Workshop Signal Proc. Adv. Wireless Commun.*, Paris, France, pp. 101–104.

Amari, S., S. C. Douglas, A. Cichocki, and H. H. Yang, July 1997c, "Novel on-line adaptive learning algorithms for blind deconvolution using the natural gradient approach," *Proc. 11th IFAC Symp. Syst. Ident.*, Kitakyushu, Japan, vol. 3, pp. 1057–1062.

Bell, A. J., and T. J. Sejnowski, Nov. 1995, "An information maximization approach to blind separation and blind deconvolution," *Neural Computation*, vol. 7, pp. 1129–1159.

Benveniste, A., M. Goursat, and G. Ruget, June 1980, "Robust identification of a nonminimum phase system: Blind adjustment of a linear equalizer in data communications," *IEEE Trans. Automat. Contr.*, vol. 25, pp. 385–399.

Cardoso, J.-F., Apr. 1997, "Infomax and maximum likelihood in source separation," *IEEE Signal Processing Lett.*, vol. 4, pp. 112–114.

Cardoso, J.-F., Oct. 1998, "Blind signal separation: Statistical principles," *Proc. IEEE*, vol. 86, pp. 2009–2025.

Cardoso, J.-F., and B. Laheld, Dec. 1996, "Equivariant adaptive source separation," *IEEE Trans. Signal Processing*, vol. 44, pp. 3017–3030.

Cichocki, A., R. Unbehauen, and E. Rummert, Aug. 1994, "Robust learning algorithm for blind separation of signals," *Electron. Lett.*, vol. 30, pp. 1386–1387.

Comon, P., Apr. 1994, "Independent component analysis: A new concept?" *Signal Processing*, vol. 36, pp. 287–314.

Delfosse, N., and P. Loubaton, July 1995, "Adaptive blind separation of independent sources: a deflation approach," *Signal Processing*, vol. 45, pp. 59–83.

Donoho, D. L., 1981, "On minimum entropy deconvolution," in D. F. Findley, ed., *Applied Time Series Analysis II*, (New York: Academic Press) pp. 565–608.

Douglas, S. C., A. Cichocki, and S. Amari, 1996, "Fast-convergence filtered regressor algorithms for blind equalisation," *Electron. Lett.*, vol. 32, pp. 2114–2115. Nov. 7, 1996.

Douglas, S. C., A. Cichocki, and S. Amari, Sept. 1997, "Multichannel blind separation and deconvolution of sources with arbitrary distributions," *Proc. IEEE Workshop Neural Networks Signal Processing*, Amelia Island, FL, pp. 436–445.

Douglas, S. C., and S.-Y. Kung, Aug. 1998a, "KuicNet algorithms for blind deconvolution," *Proc. IEEE Workshop Neural Networks Signal Processing*, Cambridge, UK, pp. 3–12.

Douglas, S. C., and S.-Y. Kung, Nov. 1998b, "Design of estimation/deflation approaches to independent component analysis," *Proc. 32nd Asilomar Conf. Signals, Syst., Comput.*, Pacific Grove, CA, vol. 1, pp. 707–711.

Douglas, S. C., S. Amari, and S. Y. Kung, Sept. 1998, "Gradient adaptation under unit-norm constraints," *Proc. IEEE Workshop Statistical Signal and Array Processing*, Portland, OR, pp. 144–147.

Douglas, S. C., S. Amari, and S.-Y. Kung, Mar. 1999a, "Adaptive paraunitary filter banks for spatio-temporal principal and minor subspace analysis," *Proc. IEEE Int. Conf. Acoust., Speech, Signal Processing*, Phoenix, AZ, vol. 2, pp. 1089–1092.

Douglas, S. C., A. Cichocki, and S. Amari, Apr. 1999b, "Self-whitening algorithms for adaptive equalization and deconvolution," *IEEE Trans. Signal Processing*, vol. 47, pp. 1161–1165.

Godard, D. N., Nov. 1980, "Self-recovering equalization and carrier tracking in two-dimensional data communication systems," *IEEE Trans. Commun.*, vol. 28, pp. 1867–1875.

Gray, R. M., Apr. 1977, "Toeplitz and circulant matrices: A review," *Technical Rept. No. 6504-1*, Inform. Sys. Lab., Stanford Univ., Stanford, CA.

Grellier, O., and P. Comon, Aug. 1988, "Blind separation of discrete sources," *IEEE Signal Processing Lett.*, vol. 5, pp. 212–214.

Haykin, S., ed., 1994, *Blind Deconvolution* (Englewood Cliffs, NJ: Prentice-Hall).

Haykin, S., 1996, *Adaptive Filter Theory* (Upper Saddle River, NJ: Prentice-Hall).

Honig, M., U. Madhow, and S. Verdu, July 1995, "Blind adaptive multiuser detection," *IEEE Trans. Inform. Theory*, vol. 41, pp. 944–960.

Hyvärinen, A., and E. Oja, Feb. 1998, "Independent component analysis by general non-linear Hebbian-like learning rules," *Signal Processing*, vol. 63, pp. 301–313.

Jain, A. K., 1988, *Fundamentals of Digital Image Processing* (Englewood Cliffs, NJ: Prentice-Hall).

Kung, S.-Y., and C. Mejuto, May 1998, "Extraction of independent components from hybrid mixture: KuicNet learning algorithm and applications," *Proc. IEEE Int. Conf. Acoust., Speech, Signal Processing*, Seattle, WA, vol. 2, pp. 1209–1212.

Lambert, R. H., May 1996, "Multichannel blind deconvolution: FIR matrix algebra and separation of multipath mixtures," Ph.D. dissertation, University of Southern California, Los Angeles, CA.

Lambert, R. H., and A. J. Bell, Apr. 1997, "Blind separation of multiple speakers in a multipath environment," *Proc. IEEE Int. Conf. Acoust., Speech, Signal Processing*, Munich, Germany, vol. 1, pp. 423–426.

Lucky, R. W., Feb. 1966, "Techniques for adaptive equalization of digital communication systems," *Bell Sys. Tech. J.*, vol. 45, pp. 255–286.

Makeig, S., T.-P. Jung, A. J. Bell, D. Ghahremani, and T. J. Sejnowski, 1997, "Blind separation of auditory event-related brain responses into independent components," *Proc. Nat. Acad. Sci.*, vol. 94, pp. 10979–10984.

McKeown, M. J., S. Makeig, G. G. Brown, T.-P. Jung, S. Kindermann, A. J. Bell, and T. J. Sejnowski, 1998, "Analysis of MRI by blind separation into independent spatial components," *Human Brain Mapping 6*, pp. 160–188.

Paulraj, A. J., and C. B. Papadias, Nov. 1997, "Space-time processing for wireless communications," *IEEE Signal Processing Mag.*, vol. 14, no. 6, pp. 49–83.

Pham, D.-T., P. Garrat, and C. Jutten, 1992, "Separation of a mixture of independent sources through a maximum likelihood approach," *Proc. EUSIPCO*, pp. 771–774.

Qureshi, S. U. H., Sept. 1985, "Adaptive equalization," *Proc. IEEE*, vol. 73, pp. 1349–1387.

Sato, Y., June 1975, "Two extensional applications of the zero-forcing equalization method," *IEEE Trans. Commun.*, vol. 23, pp. 684–687.

Torkkola, K., Sept. 1996, "Blind separation of convolved sources based on information maximization," *IEEE Workshop Neural Networks Signal Processing*, Kyoto, Japan, pp. 423–432.

Treichler, J. R., and B. G. Agee, Apr. 1983, "A new approach to multipath correction of constant modulus signals," *IEEE Trans. Acoust., Speech, Signal Processing*, vol. 31, pp. 349–372.

Treichler, J. R., M. G. Larimore, and J. C. Harp, Oct. 1998, "Practical blind demodulators for high-order QAM signals," *Proc. IEEE*, vol. 86, pp. 1907–1926.

4

BLIND SEPARATION OF INDEPENDENT SOURCES BASED ON MULTIUSER KURTOSIS OPTIMIZATION CRITERIA

Constantinos B. Papadias

ABSTRACT

The problem of blind source separation (BSS) of mutually independent independently identically distributed (i.i.d.) sources is studied from an identifiability perspective in this chapter. Motivated by the duality between the BSS and the blind equalization (BE) problems, an attempt was made to draw a parallel between the identifiability results that Shalvi and Weinstein presented in 1990 for BE and similar results that can be obtained in the BSS context. Shalvi and Weinstein studied the single-user BE problem, wherein a white input is transmitted through a linear (convolutive) channel whose output is subsequently filtered by a linear equalizer. In their seminal 1990 paper they showed that a set of necessary and sufficent conditions on the output of the equalizer exist that guarantee the perfect recovery of the input. These conditions involve second- and fourth-order cumulants of the output process, namely

Unsupervised Adaptive Filtering, Volume II, Edited by Simon Haykin.
ISBN 0-471-37941-7 © 2000 John Wiley & Sons, Inc.

its autocorrelation and *kurtosis* functions, respectively. The more general problem of linear convolutive mixtures of channels is treated here, assuming again that each source signal is i.i.d. and that they are all mutually independent and share the same non-Gaussian distribution. It is shown that, under these assumptions, a similar set of conditions exist that are necessary and sufficient for the recovery of all the input signals. In this chapter these conditions as well as a number of optimization criteria for BSS that stem from them are presented. Corresponding adaptive filtering algorithms, mainly of the stochastic-gradient type, that attempt to optimize adaptively these higher-order BSS criteria are then presented, and their performance is discussed.

4.1 INTRODUCTION

As mentioned in previous chapters in both volumes of this work, there is an intimate relationship between the BE and BSS methods. The former methods deal with the output-based recovery of a single-user signal that is transmitted through a linear dispersive channel and is thus received in the presence of intersymbol interference (ISI). The latter methods deal typically with the simultaneous recovery of a number of often independent source (user) signals that are transmitted through a linear instantaneous mixture channel. Each of the received signals then includes a linear combination of the other transmitted signals and is therefore corrupted by interuser interference (IUI). This scenario can arize, for example, in code-division multiple-access (CDMA) or space-division multiple-access communication systems. In the more general case, the signals can be received in the presence of both ISI and IUI, in which case they can be modeled as the outputs of linear dispersive matrix channels (also called "convolutive mixture channels").

The majority of blind deconvolution techniques are derived through the construction of statistical criteria that are based on the receiver outputs and typically reflect some known structural properties of the transmitted signals. As mentioned throughout both volumes of the book, a rich variety of blind recovery techniques exists in the field of blind source separation. These include independent component techniques (also called contrasts) (Comon 1994, 1996), minimum output entropy (Donoho 1981), deflation (Delfosse and Loubaton 1995), information theoretical, maximum likelihood as well as other higher-order (Swami et al. 1994; Shamsunder and Giannakis 1994; Cardoso and Laheld 1996; Yellin and Weinstein 1994; van der Veen and Paulraj 1996, Papadias and Paulraj 1997a, Cardoso 1998), and second-order (Slock 1994; Fijalkow and

Loubaton 1995) techniques. Similarly, different types of criteria have been proposed over the past 20 years for blind equalization.

The search for efficient BE criteria has been paralleled by the search for conditions that guarantee perfect blind recovery. A first step in this direction for BE was made by Benveniste et al. (1980). In that work, they showed that in the case of a single-user independently identically distributed (i.i.d.) input and a linear single-input–single-output (SISO) channel, a sufficient condition for equalization is that the equalizer output statistical distribution matches the corresponding input distribution. Despite its significance, this result was of limited practical applicability, as the output statistical distribution involves its moments of all orders. A stronger result, again for the case of SISO single-user channels with i.i.d. inputs was given by Shalvi and Weinstein (1990). In that work they derived a simple necessary and sufficient condition that the equalizer output should possess in order to guarantee perfect recovery of the input (up to an unknown delay and phase rotation). The condition is the maximization of the magnitude of the output *kurtosis* under a constant power constraint. The importance of this result is that the condition is necessary and sufficient, hence it can be used as a guideline for the design of BE optimization criteria. Moreover, it helped explain the good behavior of the popular Godard [constant-modulus (CM)] algorithm (Godard 1980; Treichler and Agee 1983) in the case of sub-Gaussian (including non-CM) inputs.

In the case of BSS, we are interested in investigating the existence of conditions that are necessary and sufficient for the recovery of all the source signals. We will assume that all the sources are i.i.d., mutually independent, and share the same non-Gaussian distribution. This assumption is made for simplification; however, it may well represent a number of source-separation paradigms (such as CDMA communication systems wherein all users use the same format for transmission). Under the previous assumptions, it turns out that, similarly to the single-user BE case a set of conditions necessary and sufficient for the recovery of all the transmitted signals exists and again involves each signal's kurtosis, as well as the autocorrelations and cross-correlations between the different source signals.

Based on these conditions, different optimization criteria for BSS can be proposed. Such criteria were recently derived by several authors and include, among other things, the single-stage kurtosis-based constrained criteria proposed in Papadias (1998), the (unconstrained) multiuser constant-modulus algorithm (CMA) (MU-CMA) criteria proposed in Papadias and Paulraj (1997a), Castedo et al. (1997), and Touzni and Fijalkow (1997), and modifications thereof proposed in Romano et al.

(1999) and Lambotharan and Chambers (1999), the multi-stage CM criteria proposed in Gooch and Lundel (1986), Shynk and Gooch (1996), Tugnait (1997), and the multistage kurtosis-based criteria proposed in Inouye (1998).

In this chapter we will attempt to present in a unified way the aforementioned identifiability conditions and corresponding constrained and unconstrained algorithms for both single-user and multiuser blind deconvolution of i.i.d. signals. A general formulation of the considered blind deconvolution problem is presented in Section 4.2. The different source signals are assumed to be i.i.d. and mutually independent, whereas the channel and receiver filters are both assumed to be linear. In Section 4.3 we review the special case of a single-source signal (single-user BE), presenting some of the important results that Shalvi and Weinstein (1990) derived. These include the necessary and sufficient condition for the BE of a single white non-Gaussian input as well as the related single-user constrained kurtosis and unconstrained CMA algorithms. Next, we extend these results to the multiuser case. In Section 4.4 we derive a set of necessary and sufficient conditions for perfect multiuser recovery in the case of linear convolutive mixtures. In Section 4.5 we present the multiuser constant-modulus (MU-CM) approach for the derivation of unconstrained source-separation algorithms for sub-Gaussian inputs. We focus on the recently proposed MU-CMA algorithm and discuss its performance. Similarly, Section 4.6 is devoted to constrained criteria and algorithms stemming from the conditions of Section 4.4: we present the so-called multiuser kurtosis (MUK) maximization criterion for BSS, which can be seen as an extension of the single-user constrained kurtosis criterion of Shalvi and Weinstein (1990). We derive an algorithmic implementation of this criterion (which we call the MUK algorithm for BSS) and discuss its performance. Section 4.7 contains some computer-simulation results of the presented techniques, which show their ability to blindly separate a number of signals that are corrupted by either IUI only (instantaneous mixture) or both IUI and ISI (convolutive mixture). Finally, Section 4.8 contains the chapter's conclusions.

4.2 PROBLEM FORMULATION AND ASSUMPTIONS

We are interested in the problem of blindly separating p source signals that are received in the presence of linear interference. We denote each of the transmitted source (user) data sequences by $a_i(k)$, $i = 1, \ldots, p$, where k is the discrete time index and we assume that each of them is a zero-mean i.i.d. discrete stochastic process. We also assume that all the source sequences share the same statistical distribution and that they are mutually independent:

$$E(a_i(k)a_j^*(k-l)) = \sigma_a^2 \delta_{ij}\delta_{kl} \tag{4.1}$$

where $*$ denotes the complex conjugate, σ_a^2 is the variance of any $\{a_i(k)\}$ and δ_{ij} is the Kronecker delta. As mentioned earlier, this scenario may well model a number of applications in communications.

Linear interference may consist of both ISI (self-interference) and IUI (interference from other users). Assuming a p-input/m-output channel, the ith channel output can be expressed (after downconversion and sampling at the symbol rate) as

$$y_i(k) = \sum_{j=1}^{p} H_{ij}\mathcal{A}_j(k) + n_i(k), \qquad i = 1, \ldots, m \tag{4.2}$$

The m channel outputs may correspond to different antenna elements in a communication application [alternatively, they may represent different sampling phases, or both sampling phases and antenna elements—see Paulraj et al. (1997)]. Here $H_{ij} = [H_{ij}(0) \cdots H_{ij}(N-1)]$ is the impulse response of the linear channel linking the jth source to the ith channel output, which is assumed to be linear [finite-duration impulse-response (FIR)] time-invariant and contains N elements; $\mathcal{A}_j(k) = [a_j(k) \cdots a_j(k-N+1)]^T$ is a vector of past input data samples corresponding to the jth user, and T denotes transpose of a matrix or vector. For a given user j, the product $H_{ij}\mathcal{A}_j(k)$ represents the total signal plus self-interference (ISI) that user j contributes to the ith channel output. On the other hand, the terms $H_{il}\mathcal{A}_l(k), l \neq j$, represent the interference that other users (IUI) contribute to the ith channel output. Finally, $n_i(k)$ represents an additive noise term present at the ith channel output (which may include thermal, measurement, and modeling components). Assuming the receiver to be linear, it produces at each of its p outputs a linear combination of several time-shifted and weighted replicas of the channel outputs $y_i(k)$. Then each of the p receiver outputs can be written as

$$z_j(k) = \sum_{i=1}^{m} W_{ij}^T Y_i(k) = W_j^T Y(k), \qquad j = 1, \ldots, p \tag{4.3}$$

where $Y_i(k) = [y_i(k) \cdots y_i(k-M+1)]^T$ is a vector of M past samples of the ith channel output; and $W_{ij} = [W_{ij}(0) \cdots W_{ij}(M-1)]^T$ is the "equalizer" linking $Y_i(k)$ to the jth receiver output. Here M represents the length of each equalizer (in symbol periods) and W_j, $Y(k)$ are defined as $W_j = [W_{1j}^T \cdots W_{mj}^T]^T$, $Y(k) = [Y_1^T(k) \cdots Y_m^T(k)]^T$, respectively. We can now express each $z_j(k)$ in terms of the input source signals as follows

$$z_j(k) = W_j^T Y(k) = W_j^T \mathscr{H} A(k) + W_j^T N(k) \tag{4.4}$$

where

$$\mathscr{H} = \begin{bmatrix} \boxed{H_1} & 0 & \cdots & 0 \\ 0 & \boxed{H_1} & \ddots & \vdots \\ \vdots & \ddots & \ddots & 0 \\ 0 & \cdots & 0 & \boxed{H_1} \\ \vdots & \vdots & \vdots & \vdots \\ \boxed{H_m} & 0 & \cdots & 0 \\ 0 & \boxed{H_m} & \ddots & \vdots \\ \vdots & \ddots & \ddots & 0 \\ 0 & \cdots & 0 & \boxed{H_m} \end{bmatrix} \tag{4.5}$$

with

$$H_i = [H_{i1}(0) \cdots H_{ip}(0) \cdots H_{i1}(N-1) \cdots H_{ip}(N-1)], \qquad i \in \{1, \ldots, m\}$$

$$A(k) = [a_1(k) \cdots a_p(k) \cdots a_1(k-N-M+2) \cdots a_p(k-N-M+2)]^T$$

and $N(k) = [N_1^T(k) \cdots N_m^T(k)]^T$, $N_i(k) = [n_i(k) \cdots n_i(k-M+1)]^T$. The "0" elements in $\mathscr{H}$ are each a $1 \times p$ vector of zeros, resulting in the size of $\mathscr{H}$ being $mM \times p(N+M-1)$. Notice that, if M is chosen so that $mM > p(N+M-1)$, $\mathscr{H}$ will be a "tall" matrix (more rows than columns). Defining $G_j^T = W_j^T \mathscr{H}$, we can express each of the p receiver outputs as

$$z_j(k) = G_j^T A(k) + W_j^T N(k) \tag{4.6}$$

where G_j now represents the global impulse response between the p input sources and the jth receiver output, and is the convolution of the channel matrix with the equalizer intended for user j, as expressed through

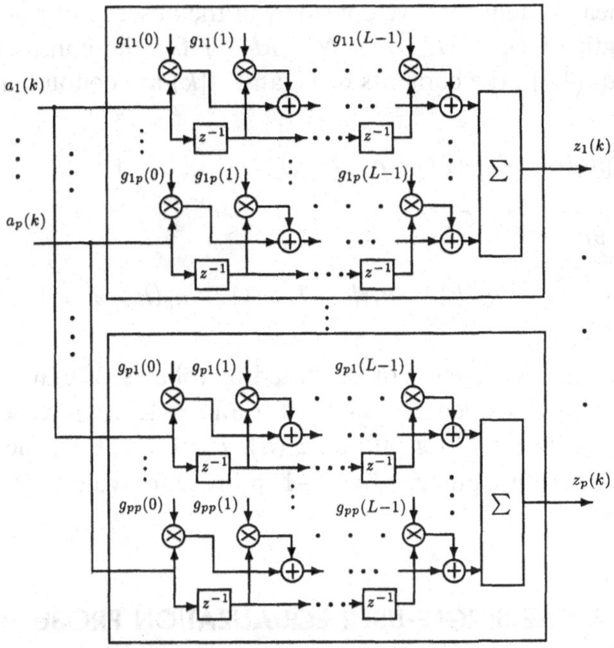

Figure 4.1 A multiuser multiple-input–multiple output (MIMO) equalization/source-separation setup.

$G_j^T = W_j^T \mathscr{H}$. Equation (4.6) can be also written in matrix form as

$$\mathbf{z}(k) = \mathbf{G}^T A(k) + \mathbf{n}(k) \tag{4.7}$$

where

$$\mathbf{G} = [G_1 \cdots G_p] \tag{4.8}$$

$\mathbf{z}(k) = [z_1(k) \cdots z_p(k)]^T$ and $\mathbf{n}(k) = \mathbf{W}^T N(k)$, the $mM \times p$ equalizer matrix $\mathbf{W}$ being defined as

$$\mathbf{W} = [W_1 \cdots W_p] \tag{4.9}$$

Notice that due to the preceding assumptions, A is stationary; however, in order to obtain $A(k-1), A(k)$ must be shifted by p positions downward. It will become clear in the rest of the chapter that there are certain advantages in parameterizing the system in terms of its overall input/output response $\mathbf{G}$, instead of the channel and equalizer parameters. The way that $\mathbf{G}$ is linked to $\mathscr{H}$ and $\mathbf{W}$ then becomes of somewhat secondary importance. Figure 4.1 depicts a representation of the overall p-input/

p-output linear system that we consider, in the absence of noise. Denoting the length of G_j by L_p^n [$L = N + M - 1$ for the considered signal model of Eq. (4.7)], the contents of G_j and $A(k)$ are denoted as

$$\begin{aligned} G_j &= [g_{1j}(0) \cdots g_{pj}(0) \cdots g_{1j}(L-1) \cdots g_{pj}(L-1)]^T \\ &\equiv [g_{j1} \cdots g_{jLp}]^T \\ A(k) &= [a_1(k) \cdots a_p(k) \cdots a_1(k-L+1) \cdots a_p(k-L+1)]^T \end{aligned} \tag{4.10}$$

The problem that we treat can be stated as follows: We aim at the retrieval of the input sequences $\{a_j(k)\}$, for all $i = 1, \ldots, p$, based only on the statistics of the equalizer outputs $z_j(k)$, $j = 1, \ldots, p$. In the following we will denote the imaginary j by $\sqrt{-1}$ in order to avoid confusion with other indices.

4.3 REVIEW: THE SINGLE-USER EQUALIZATION PROBLEM

The single-user BE problem can be seen as a special case of the multiuser problem stated in the previous section. In this case, we assume a single white i.i.d. input $a(k)$, which is transmitted through a channel H and is subsequently filtered by an equalizer W. Assuming a symbol-rate receiver, the single-channel output takes the form [similar to Eq. (4.2)]

$$y(k) = H\mathcal{A}(k) + n(k) \tag{4.11}$$

where H is $1 \times N$ and $\mathcal{A}(k) = [a(k) \cdots a(k-N+1)]^T$. The equalizer output is then simply given by

$$z(k) = W^T Y(k) \tag{4.12}$$

where the length of the equalizer W is M and $Y(k) = [y(k) \cdots y(k-M+1)]^T$. Similar to Eq. (4.6), we can also express $z(k)$ as

$$z(k) = G^T A(k) + W^T N(k) \tag{4.13}$$

where $G^T = W^T \mathcal{H}$, $A(k) = [a(k), \ldots, a(k-N-M+2)]^T$ $N(k) = [n(k) \cdots n(k-M+1)]^T$, and $\mathcal{H}$ is now defined as

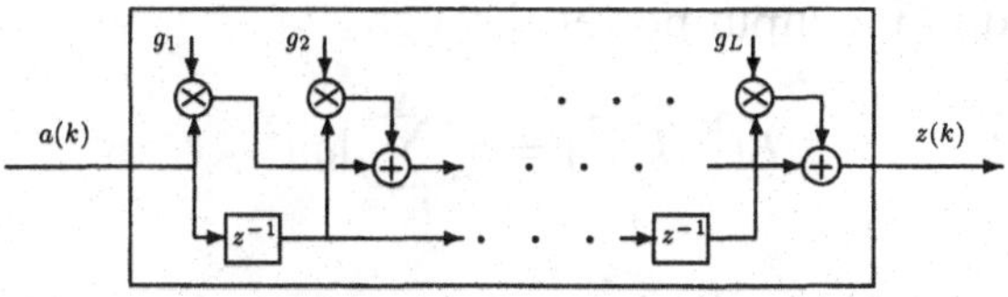

Figure 4.2 A single-user equalization setup.

$$\mathcal{H} = \begin{bmatrix} \boxed{H} & 0 & \cdots & 0 \\ 0 & \boxed{H} & \ddots & \vdots \\ \vdots & \ddots & \ddots & 0 \\ 0 & \cdots & 0 & \boxed{H} \end{bmatrix} \tag{4.14}$$

The "0" elements in Eq. (4.14) are now scalar. Notice that the channel matrix $\mathcal{H}$ is now $M \times (M + N - 1)$, hence it always has more columns than rows—a wide matrix [only asymptotically for $M \to \infty$ may $\mathcal{H}$ approach a full column rank (square) matrix]. Denoting $L = N + M - 1$, the signal model in the single-user case is, in the absence of noise,

$$z(k) = G^T A(k) \tag{4.15}$$

where $G = [g_1 \cdots g_L]^T$ and $A(k) = [a(k) \cdots a(k - L + 1)]^T$. The setup of the single-user BE problem (again in the absence of noise) considered is depicted in Fig. 4.2. In Shalvi and Weinstein (1990) the authors showed that, when the input is i.i.d., two statistical quantities of the equalizer output are of particular importance for BE, namely, its second-order moment [or power—$E|z(k)|^2$] and its *kurtosis*. The (unnormalized) kurtosis of a stationary stochastic process $\{x\}$ is a fourth-order cumulant defined as

$$K(x) = E(|x|^4) - 2E^2(|x|^2) - |E(x^2)|^2 \tag{4.16}$$

The importance of these second- and fourth-order cumulants of the output $z(k)$ can be seen if one tries to express them in terms of the corresponding input quantities. Based on Eq. (4.15), and keeping in mind that $a(k)$ is white and nonzero, that is,

$$E(a(k)a^*(k - l)) = \sigma_a^2 \delta_{kl} \tag{4.17}$$

it is straighforward to show (Shalvi and Weinstein 1990) that the output power is linked to the input power through

$$E(|z(k)|^2) = \sigma_a^2 \sum_{l \in \mathscr{P}} |g_l|^2 \tag{4.18}$$

where $\mathscr{P} = \{1, \ldots, L\}$. Similarly, the kurtosis of $z(k)$ can be expressed as a function of the input kurtosis

$$K(z(k)) = K_a \sum_{l \in \mathscr{P}} |g_l|^4 \tag{4.19}$$

where $K_a = K(a)$. From Eqs. (4.18) and (4.19), it was noticed in Shalvi and Weinstein (1990) that, if $E|z(k)|^2 = \sigma_a^2$, then

$$|K(z(k))| \leq |K_a| \tag{4.20}$$

which stems from the fact that

$$\sum_{l \in \mathscr{P}} |g_l|^4 \leq \left(\sum_{l \in \mathscr{P}} |g_l|^2 \right)^2 \tag{4.21}$$

Another important consequence of Eq. (4.21) is that equality in Eq. (4.21) holds *if and only if* G has a unique nonzero element g_l of unit magnitude for some $l \in \mathscr{P}$. Using these findings, Shalvi and Weinstein formulated the following theorem on the blind identifiability of a white input based on the output of a linear channel/equalizer cascade.

Theorem 1. A necessary and sufficient condition for the equalization (up to a scalar-phase ambiguity) of a zero-mean i.i.d. complex random sequence at the output of a SISO linear channel/equalizer cascade is that the following two conditions be satisfied simultaneously:

$$\begin{cases} |K(z(k))| = |K_a| \\ E|z(k)|^2 = \sigma_a^2 \end{cases} \tag{4.22}$$

As mentioned in the chapter's introduction, the importance of this condition is that it is necessary and sufficient, that is, *any* method that is used to achieve equalization will have to satisfy Eq. (4.22) (in the absence of noise). Hence Eq. (4.22) can be used as a guideline for the design of single-user BE optimization criteria. Notice also that in light of the way this single-user problem was formulated, Theorem 1 implies that, with an FIR channel, the equalizer has to be of infinite length in order to satisfy equality in Eq. (4.21) (in other words $M, L \to \infty$). With a finite-length equalizer (L finite), perfect equalization in the absence of noise is not possible when the equalizer is SISO; (however, a good approximation can be achieved with a "long enough" equalizer). This is due to the fact that, as mentioned before, $\mathcal{H}$ in Eq. (4.14) is a "wide" matrix, hence it has no "left-hand-side inverse" (this limitation can be overcome if a fractionally spaced [multiple-input-single-output (MISO)] equalizer is used instead [see Papadias and Slock (1999)]).

In Shalvi and Weinstein (1990), a number of interesting equalization criteria were proposed and analyzed for single-user BE. We now review some of those results.

4.3.1 Constrained Criteria

Based on the condition Eq. (4.22), the following constrained optimization problem was suggested for BE in Shalvi and Weinstein (1990)

$$\begin{cases} \max\limits_{G} & \mathcal{F}_{\mathrm{SW}}(G) = |K(z)| \\ \text{subject to:} & G^H G = 1 \end{cases} \tag{4.23}$$

where H denotes Hermitian (complex conjugate) transpose and it has been assumed, for simplicity that $\sigma_a^2 = 1$. By constructing the Lagrangian function corresponding to the constrained criterion Eq. (4.23), the stationary points of $F_{\mathrm{SW}}(G)$ on $G^H G = 1$ were found in Shalvi and Weinstein (1990), and are given by all vectors G such that

$$\begin{aligned} r_i^2 &= 1/P, \qquad i \in \{I_P\} \\ r_i^2 &= 0, \qquad i \notin \{I_P\} \end{aligned} \tag{4.24}$$

where $r_i = |g_i|$ is the magnitude of the ith element of G and I_P is any P-element subset of $\mathcal{P}$. In other words, the stationary points of Eq. (4.23) are given by the sets of unit-norm vectors whose P nonzero elements are all equimagnitude. Moreover, it was shown in Shalvi and Weinstein (1990) that, apart from the global maxima corresponding to $P = 1$ in Eq. (4.24), all other solutions are unstable saddle points of $\mathcal{F}_{\mathrm{SW}}(G)$. This result is summarized in the following theorem.

Theorem 2. The only stable stationary points of the constrained single-user kurtosis cost function $\mathcal{F}_{\mathrm{SW}}(G)$ in Eq. (4.23) (in the absence of noise) are its global maxima given by

$$G = e^{\sqrt{-1}\phi}[\cdots \; 0 \; 1 \; 0 \; \cdots]^T \tag{4.25}$$

which correspond to the perfect (zero-forcing) recovery of the transmitted input $a(k)$, up to an arbitrary delay and scalar phase rotation ϕ.

The result of Theorem 2 is important, as it implies that a blind constrained algorithm designed to satisfy the criterion Eq. (4.24) will be globally convergent to a setting that recovers the input, up to a delay and phase rotation.[1] Such a stochastic-gradient-type algorithm was derived in Shalvi and Weinstein (1990). Assuming the input to be complex and symmetrical $[E(a(k))^2 = 0]$, the algorithm is given by the following two-step update procedure:

$$\begin{aligned} W'(k+1) &= W(k) + \mu\,\text{sign}(K_a)(|z(k)|^2 z(k))\,Y^*(k) \\ W(k+1) &= W'(k+1)/\|W'(k+1)\| \end{aligned} \tag{4.26}$$

where $\|X\| = \sqrt{X^H X}$ is the Euclidean norm of a column vector X. A somewhat more involved algorithm for the case $E(a(k))^2 \neq 0$ was also derived in Shalvi and Weinstein (1990). In both cases, it is important to emphasize that these single-user constrained kurtosis maximization algorithms require the prewhitening of the channel output $y(k)$. This is required in order to guarantee that the property $W^H W = 1$, which the algorithm tries to enforce, will reflect to the required property $G^H G = 1$. Since $G^T = W^T \mathscr{H}$, it is easy to see from Eq. (4.14) that this requires $\mathscr{H}$ to be unitary ($\mathscr{H}\mathscr{H}^H = \mathbf{I}$). This condition, in the frequency domain, corresponds to

$$|H(f)| = 1 \tag{4.27}$$

which means that the channel must be flat in frequency. Given the assumed whiteness of $a(k)$, this results in $y(k)$ having to be white too. Whitening of $y(k)$ can be achieved either by spectral prewhitening, or by time-domain techniques such as linear prediction.

Besides the strong feature of global convergence (after prewhitening), the algorithm given in Eq. (4.26) has another important property: it applies equally to both sub-Gaussian ($K_a < 0$) and to super-Gaussian ($K_a > 0$) inputs. This is an advantage over most single-user BE methods, which are limited in that they can only cope with sub-Gaussian signals. While super-Gaussian distributions are not used in typical communication systems, it is possible that, due to spectral shaping, the transmitted

[1] We emphasize again that, strictly speaking, an infinite-length equalizer $[\mathscr{P} = \mathbb{Z}]$ is required for G in Eq. (4.25) to be achievable for the considered SISO model in Eq. (4.11).

signal approaches a super-Gaussian distribution. Of course, it is clear from the study just discussed that all kurtosis-based methods will fail in the case of Gaussian inputs (in which case $K_a = 0$).

4.3.2 Unconstrained Criteria

An unconstrained counterpart to the single-user kurtosis maximization criterion, Eq. (4.23), is the well-known CMA 2-2 criterion,

$$\min_{W} J_{\text{CM}}(W) = E(|z(k)|^2 - 1)^2 \tag{4.28}$$

where we have assumed for simplicity that $E|a(k)|^4/E|a(k)|^2 = 1$. The stochastic-gradient optimization of Eq. (4.28) is achieved through the popular CMA algorithm:

$$W(k+1) = W(k) + \mu(|z(k)|^2 - 1)z(k)Y^*(k) \tag{4.29}$$

While the CMA cost function in Eq. (4.28) was derived in Godard (1980) and Treichler and Agee (1983) in a somewhat ad hoc way, it was shown in Shalvi and Weinstein (1990) that it is closely linked to conditions (4.22) and to the criterion, Eq. (4.23). Specifically, it turns out that, in the sub-Guassian case ($K_a < 0$), the minimization of J_{CM} in Eq. (4.28) is equivalent to the maximization of $K(z)/K_a + \Delta(z)$, where $\Delta(z)$ is a constant function for fixed $E|z|^2$. As a result, it turns out that, if $K_a < 0$, the minimization of $J_{\text{CM}}(G)$ coincides with the maximization of $J_{\text{SW}}(G)$ over $G^H G = 1$. Hence, the CMA criterion can be seen as a special case of the kurtosis maximization criterion, Eq. (4.23) in the case of sub-Gaussian inputs! This reinforces the value of the CMA algorithm, as it no longer corresponds to an ad hoc "constant modulus" criterion, but rather to the satisfaction of the necessary and sufficient conditions (4.22) for all sub-Gaussian inputs.

As implied above, it was shown in Shalvi and Weinstein (1990) that a one-to-one correspondence between the stationary points of the two criteria exists. This results in the CMA being globally convergent to an ideal Dirac setting in the absence of noise (again, if the SISO equalizer W has infinite length).[2] We summarize these results in the following theorem.

[2] The fact that zero-forcing equalization of FIR channels is possible with FIR equalizers if the channel is oversampled (see (Papadias and Slock 1999) and references therein) is behind the proof that the CMA is globally convergent even when the equalizer has finite length, provided that it is fractionally spaced (Li and Ding 1996; Fijalkow et al. 1997).

Theorem 3. The CMA cost function in Eq. (4.28) has no undesired local minima in the G domain, provided that the input is sub-Gaussian ($K_a < 0$), G is allowed to be infinite length, and no noise is present.

As a result, in these cases, the CMA 2-2 algorithm, Eq. (4.29), will be globally convergent to an ideal Dirac-like solution of the type shown in Eq. (4.25). Moreover, the CMA criterion, Eq. (4.28), which is unconstrained, can be seen as a special case of the constrained criterion, Eq. (4.23), for the case of sub-Gaussian inputs.

It should be noted at this point that it was known before that the CMA 2-2 is globally convergent (Foschini 1985) (where the inputs were implicitly assumed to be symmetrical and sub-Gaussian). It is the work in Shalvi and Weinstein (1990) though that made it clear why this is true for all sub-Gaussian inputs, and that established the link between the CMA and the single-user Kurtosis conditions, Eq. (4.22).

In the rest of the chapter, we extend these results to the BSS problem (more than one user in the system).

4.4 NECESSARY AND SUFFICIENT CONDITIONS FOR BSS

We now go back to our original problem stated in Section 4.2. As in the BE case, we allow each transmitted signal to be recovered up to a time delay and unitary scalar ambiguity. Therefore, blind recovery will be achieved if, after suitable reordering of the equalizer outputs, the following condition holds:

$$z_j(k) = e^{\sqrt{-1}\phi_j} a_j(k - \delta_j) \tag{4.30}$$

for some $\phi_j \in [0, 2\pi)$, $\delta_j \in \mathbb{Z}$ and all $j \in \{1, \ldots, p\}$.

We will attempt to follow the same steps as in Section 4.3 in order to study the identifiability problem. In analogy to Eq. (4.18), we can now write for each receiver output,

$$E|z_j(k)|^2 = \sigma_a^2 \sum_{l=1}^{Lp} |g_{jl}|^2, \qquad j = 1, \ldots, p \tag{4.31}$$

and, similar to Eq. (4.19),

$$K(z_j(k)) = K_a \sum_{l=1}^{Lp} |g_{jl}|^4, \qquad j = 1, \ldots, p \tag{4.32}$$

where we now recall that σ_a^2 and K_a are the variance and the kurtosis of *each* $a_i(k)$ (since they all have the same distribution), respectively:

$$\begin{aligned} K_a &= K(a_i) = E(|a_i|^4) - 2E^2(|a_i|^2) - |E(a_i^2)|^2 \\ \sigma_a^2 &= E(|a_i|^2) \end{aligned} \tag{4.33}$$

Based on Eqs. (4.31) and (4.32), we can now formulate the following theorem [see also Papadias and Paulraj (1997b)][3]:

Theorem 4. If each $a_i(k), i = 1, \ldots, p$ is an i.i.d. zero-mean sequence, $\{a_i(k)\}$, $\{a_j(k)\}$ are statistically independent for $i \neq j$ and share the same statistical properties, then the following set of conditions is necessary and sufficient for the recovery of all the transmitted signals at the equalizer outputs:

$$\begin{aligned} &(\text{C1}) \quad |K(z_j(k))| = |K_a|, \qquad j = 1, \ldots, p \\ &(\text{C2}) \quad E|z_j(k)|^2 = \sigma_a^2, \qquad j = 1, \ldots, p \\ &(\text{C3}) \quad E(z_i(k)z_j^*(k-l)) = 0, \qquad i \neq j, \qquad l = -L+1, \ldots, L-1 \end{aligned}$$

Proof

Necessity. To achieve perfect recovery, Eq. (4.30) must hold, from which (C1) and (C2) follow immediately and (C3) follows since $E(a_i(k-\delta_i)a_j^*(k-\delta_j)) = 0$ for $i \neq j$.

Sufficiency. From Eq. (4.32) and (C1) we get $\sum_{l=1}^{Lp} |g_{jl}|^4 = 1$. From Eq. (4.31) and (C2) we get $\sum_{l=1}^{Lp} |g_{jl}|^2 = 1$. Therefore G_j must be of the form

$$G_j = [0 \cdots e^{\sqrt{-1}\phi_j}\ 0 \cdots 0]^T \tag{4.34}$$

where the single nonzero element can be at *any* position. Now combining (C3) with Eq. (4.6), we get:

$$G_i^H J_{lp} G_j = 0 \tag{4.35}$$

where J_l is an $L \times L$ matrix, whose elements equal 1 if they belong to the lth lower diagonal ($l > 0$) or to the $|l|$th upper diagonal ($l < 0$) and 0

[3] The author recently became aware of an independently derived proof of Theorem 4 in Inouye (1997).

elsewhere (e.g., $J_0 = \mathbf{I}$). According to Eqs. (4.34) and (4.35), if G_j's nonzero element is at position n, then G_i cannot have its nonzero element at position $n - lp$. Applying all this to $l \in \{-L+1, \ldots, L-1\}$, we get a set of G_j, $j \in \{1, \ldots, p\}$, whose nonzero elements are in positions that correspond to possibly delayed versions of the p different inputs $a_j(k)$. Therefore, after reordering, we obtain Eq. (4.30), and perfect equalization/signal separation has been achieved.

Theorem 4 can be seen as a straightforward extension of Theorem 1 to the multiuser case, under the assumptions of Section 4.2 [mutually independent (i.i.d.) sources with the same probability density function (pdf)]. While conditions (C1) and (C2) coincide, for each user, to the conditions in Eq. (4.22), condition (C3) is crucial when source separation is considered. Since all the sources are assumed to share the same statistical distribution, conditions (C1) and (C2) alone would not suffice for source separation: the individual statistical properties imposed by (C1) and (C2) carry no information about *joint* output statistics. As an example, imagine that the same exactly input, say $a_1(k)$, is recovered by both $z_1(k)$ and by $z_2(k)$. Then (C1) and (C2) would both be satisfied, but separation would have not been achieved since that not all the sources would have been recovered. The role of (C3) is to avoid such situations: in the considered example, (C3) would not be satisfied if $z_1(k) = z_2(k) = a_1(k)$. Moreover, the theorem tells us that (C1), (C2), and (C3) are not only sufficient, but necessary too. So, as with the single-user case, we expect any BSS technique that is used to separate mutually independent sources with the same distribution to be linked to these conditions. Conditions (C1), (C2), and (C3) can thus be used as guidelines for the design of source-separation criteria. Moreover, they could be used as a statistical test for the measure of success of BSS methods.

In what follows we describe two families of BSS criteria that stem from these conditions.

4.5 UNCONSTRAINED CRITERIA: THE MU-CM APPROACH

We first concentrate on unconstrained criteria that are linked to the conditions of Theorem 4 for BSS. The attractive features of the single-user CMA algorithm, Eq. (4.29), which is derived from the unconstrained criterion, Eq. (4.28), make it desirable to design similar unconstrained criteria for the BSS problem.

A first attempt in this direction was made in Treichler and Larimore (1985) and Gooch and Lundel (1986). This approach attempts the extension of the CMA algorithm, Eq. (4.29), to the multiuser case by minimizing the following cost function:

$$J_1(\mathbf{W}) = E\sum_{j=1}^{p}(|z_j|^2 - 1)^2 \tag{4.36}$$

which gives the following stochastic-gradient algorithm:

$$\mathbf{W}(k+1) = \mathbf{W}(k) - \mu Y^*(k)\mathscr{Z}(k) \tag{4.37}$$

where

$$\mathbf{Y} = [Y_1^T(k) \cdots Y_m^T(k)]^T,$$

$$\mathscr{Z}(k) = [(|z_1(k)|^2 - 1)z_1(k) \cdots (|z_p(k)|^2 - 1)z_p(k)]^T.$$

Unfortunately, the simple cost function in Eq. (4.36) has an inherent limitation: some of its global minima correspond to the retrieval of only *some* of the transmitted input signals. This happens because even if the same input signal is recovered at all outputs ($z_1(k) = \cdots = z_p(k) = a; (k)$), $J_1(\mathbf{W})$ in Eq. (4.36) will still attain its minimal value (zero). In light of the conditions of Theorem 4, this is due to the failure of $J_1(\mathbf{W})$ to impose condition (C3). As a result, the algorithm, Eq. (4.37), could converge to a global minimum that, even in the absence of noise and perfect power balance between the different sources, would fail to recover all the sources.

In order to avoid this problem, it is clear that condition (C3) has to be somehow incorporated in the BSS cost function. Such a CM-type approach was proposed in Papadias and Paulraj (1997a) and is described in the following section.

4.5.1 The MU-CMA Algorithm

By adding to $J_1(\mathbf{W})$ a penalization term that is intended to prevent the violation of condition (C3), we construct the following cost function for BSS of independent sources:

$$J_{\text{MU-CMA}}(\mathbf{W}) = E\sum_{j=1}^{p}(|z_j|^2 - 1)^2 + \alpha \sum_{l,n=1;l\neq n}^{p} \sum_{\zeta=\delta_1}^{\delta_2} |r_{ln}(\zeta)|^2 \tag{4.38}$$

where $r_{ln}(\zeta)$ is the cross-correlation function between users l and n, defined as

$$r_{ln}(\zeta) = E(z_l(k)z_n^*(k-\zeta)) \tag{4.39}$$

The cost function in Eq. (4.38) is the weighted sum of a CM term and a cross-correlation term: the CM term penalizes the deviations of the equalized signals' magnitudes from a constant modulus, whereas the cross-correlation term penalizes the correlations between them. Notice that the correlation term is directly inspired by condition (C3) of Theorem 4; α is a weighting scalar parameter; and δ_1 and δ_2 are integers that should be chosen in compliance with the channel delay spread in order to consider all the achievable delays between the p user signals (for example, if L is known, we should choose $\delta_1 = -\delta_2 = -L + 1$). The stochastic-gradient algorithm that minimizes Eq. (4.38) has the form

$$\mathbf{W}(k+1) = \mathbf{W}(k) - \mu[\Delta_1(k) \cdots \Delta_p(k)] \tag{4.40}$$

where

$$\Delta_j(k) = (|z_j(k)|^2 - 1)z_j(k)Y^*(k) + \frac{\alpha}{2} \sum_{l=1;l\neq j}^{p} \sum_{\zeta=\delta_1}^{\delta_2} \hat{r}_je(\delta)z_l(k-\zeta)Y^*(k) \tag{4.41}$$

and $\hat{r}_je(\delta)$ is an estimate of $r_je(\delta)$ based on instantaneous values or sample averaging (e.g., with a sliding or growing window). We refer to Eq. (4.40) as the multi-user constant modulus algorithm (MU-CMA) for the blind deconvolution of linear dispersive mixing channels. Notice that Eq. (4.40) reduces to the single-user CMA 2-2 (Godard 1980; Treichler and Agee 1983) for $p = 1$.

The following holds for the global minima of the cost function, Eq. (4.38).

Theorem 5. If all the assumed conditions are met, the global minima of Eq. (4.38) correspond to the recovery of all the transmitted signals $a_j(k), j = 1, \ldots, p$, provided that δ_1, δ_2 are chosen to span the channel's delay spread.

See Papadias and Paulraj (1997a) for a proof of this theorem.

According to Theorem 5, the MU-CMA circumvents the identifiability problem of the algorithm in Eq. (4.37), since all its global minima are desired. Moreover, it has been shown in Castedo et al. (1997) that, in the instantaneous mixture case, the MU-CMA has no undesired stable local minima in the case of $p = 2$ users. Extensive simulation results by several researchers indicate that the algorithm has very good convergence behavior in other cases as well (arbitrary number of sources at different power levels). Combined with its simplicity, these features make the MU-CM approach an attractive one for unconstrained BSS.

Several other unconstrained algorithms have been derived that are linked to the conditions of Theorem 4. We divide these algorithms into single-stage and multistage. Single-stage approaches attempt, similarly to the MU-CMA, to recover all the source signals simultaneously. They are typically more complex, but require a smaller number of adaptation cycles for convergence. Multistage approaches typically recover the source signals one by one: after the first signal is extracted, its effect is subtracted from the output before the second signal is sought, and so forth. These approaches are typically simpler and require less computations/cycle, but require more iterations before they converge. Especially for high numbers of users, this constraint may be prohibitive. Deflation-type algorithms (see Delfosse and Loubaton (1995)) constitute a third (intermediate) category. While these do not require the subtraction of previously detected signals, the recovery of each output is focilitated by the recovery of other outputs.

The MU-CMA was independently proposed (in the context of zero delay spread) in Castedo et al. (1997), and in Touzni and Fijalkow (1997). A single-stage least-squares extension of the MU-CMA was recently proposed in Romano et al. (1999). This approach improves the algorithm's convergence speed at the cost of higher computational complexity. An interesting deflation-type variant of the MU-CMA was recently proposed in Lambotharan and Chambers (1999). Assuming sub-Gaussian inputs and some other mild conditions, the resulting algorithm was shown to be globally convergent to a setting that recovers all the source signals (after all equalizer outputs have converged). A well-known multistage algorithm based directly on the single-user CMA, Eq. (4.29), is the multistage CMA proposed in Gooch and Lundel (1986) and Shynk and Gooch (1996). Global convergence of a similar approach was recently shown in Tugnait (1997).

We should also mention that following an approach similar to Shalvi and Weinstein (1990), it is possible to derive unconstrained criteria for the super-Gaussian case. In the following section, we concentrate on constrained criteria stemming from the conditions of Theorem 4.

4.6 CONSTRAINED CRITERIA: THE MUK APPROACH

Like the single-user criterion, Eq. (4.23), which was inspired by Theorem 1, it is possible to construct multiuser kurtosis-based constrained criteria stemming from Theorem 4 and conditions (C1), (C2), and (C3). Single-stage criteria of this type were recently proposed for the instantaneous mixture case in Cardoso and Laheld (1996), Papadias and Paulraj (1997b), and Papadias (1998), whereas both single- and multistage criteria for the dispersive mixture case were proposed in Inouye (1998).

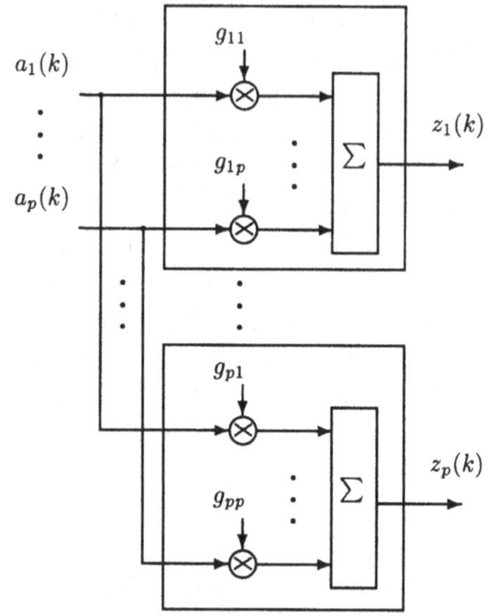

Figure 4.3 A multiuser linear-mixture setup.

Henceforth, we focus on the instantaneous-mixture case and describe a basic criterion that somewhat underlies all these works.

In the instantaneous-mixture case the system is memoryless and can be represented as shown in Fig. 4.3. A straightforward criterion for this case then can be drawn from conditions (C1), (C2), and (C3) of Theorem 4 as follows:

$$\begin{cases} \max\limits_{\mathbf{G}} & \mathscr{F}_{\text{MUK}}(\mathbf{G}) = \sum\limits_{j=1}^{p} |K(z_j)| \\ \text{subject to:} & \mathbf{G}^H\mathbf{G} = \mathbf{I} \end{cases} \tag{4.42}$$

where $\mathbf{I}$ is the $p \times p$ identity matrix. The constraint in Eq. (4.42) comes from the fact that, according to (C2) and (C3),

$$E(\mathbf{z}\mathbf{z}^H) = \sigma_a^2\mathbf{I} \tag{4.43}$$

[in Eq. (4.42) we have assumed again for simplicity that $\sigma_a^2 = 1$]. We call Eq. (4.42) the Multi-User Kurtosis (MUK) maximization criterion for BSS. Notice the similarity between this criterion and the single-user criterion, Eq. (4.23). According to Theorem 4, the global optima of the MUK criterion, Eq. (4.42), achieve blind recovery. Hence the MUK criterion has the same desired optimality of its global settings as the one expressed in Theorem 5 about the MU-CMA algorithm:

Theorem 6. The global maxima of the MUK criterion, Eq. (4.42), correspond to the recovery of all the transmitted signals $a_j(k), j = 1, \ldots, p$, up to an arbitrary phase rotation and ordering.

A similar result holds for the extension of the criterion, Eq. (4.42) to the finite ISI case [see also Inouye (1998)].

Regarding local behavior, it was shown in Papadias (1998) that, similarly to the MU-CM cost function of Eq. (4.38), the MUK criterion in Eq. (4.42) has no stable local maxima in the case of $p = 2$ users. It was also recently shown in Papadias (1999) that several classes of stationary points of Eq. (4.42) that can be analytically described for an arbitrary number of users contain no local maxima either, implying the good convergence behavior of corresponding algorithms. However, this would still leave open the possibility of stable undesired local convergence points (for $p > 2$) that have not been described analytically. On the other hand, the different algorithms that can be proposed based on the criterion (4.42) will have potentially different convergence behaviors, due especially to the way that they enforce the orthogonality constraint in their adaptation. Hence, it may be more useful to study the convergence of particular techniques, instead of the criterion itself. In the rest of the section we will present a novel adaptive algorithm that was recently derived in the spirit of the criterion in Eq. (4.42) and is especially appealing from the point of view of convergence.

4.6.1 The MUK Algorithm

We denote the $m \times p$ channel matrix by $\mathbf{H}$ and the $m \times 1$ channel output vector by $Y(k)$. The received signal model is then

$$Y(k) = \mathbf{H}A(k) + \mathbf{n}(k) \tag{4.44}$$

where $\mathbf{n}(k)$ is the $m \times 1$ vector of additive noise samples and $A(k) = [a, (k), \ldots, ap(k)]^T$. The receiver output can then be written as

$$\mathbf{z}(k) = \mathbf{W}^T(k)Y(k) = \mathbf{W}^T(k)\mathbf{H}A(k) + \mathbf{n}'(k) = \mathbf{G}^T(k)A(k) + \mathbf{n}'(k) \tag{4.45}$$

where $\mathbf{W}(k)$ and $\mathbf{G}^T(k) = \mathbf{W}^T(k)\mathbf{H}$ are the $m \times p$ receiver matrix and $p \times p$ global response matrix, respectively, and $\mathbf{n}'(k) = \mathbf{W}^T(k)\mathbf{n}(k)$ is the colored noise at the receiver output, all at time instant k.

A stochastic-gradient-type algorithm for the constrained criterion, Eq. (4.42), can be obtained as follows [Papadias (1999)]. We first compute the gradient of $\mathscr{F}_{\mathrm{MUK}}(\mathbf{G})$ with respect to $\mathbf{W}$. By writing $\mathscr{F}_{\mathrm{MUK}}(\mathbf{G})$ as

$$\begin{aligned}\mathscr{F}_{\mathrm{MUK}}(\mathbf{G}) &= \sum_{j=1}^{p} \operatorname{sign}(K(z_j))K(z_j) \\ &= \operatorname{sign}(K_a) \sum_{j=1}^{p} (E|z_j|^4 - 2E^2|z_j|^2 - |E(z_j^2)|^2)\end{aligned}$$

and assuming symmetrical inputs $(E(a^2(k) = 0))$, the gradient of $\mathscr{F}_{\mathrm{MUK}}(\mathbf{G})$ with respect to $\mathbf{W}$ equals [similar to Shalvi and Weinstein (1990)]

$$\nabla(\mathscr{F}_{\mathrm{MUK}}(\mathbf{G})) = 4 \sum_{j=1}^{p} E(|z_j(k)|^2 z_j(k) Y^*(k)) \tag{4.46}$$

We first perform an update of $\mathbf{W}(k)$ in the direction of the instantaneous gradient [dropping the expectation operator in Eq. (4.46)] as follows:

$$\mathbf{W}'(k+1) = \mathbf{W}(k) + \mu \operatorname{sign}(K_a) Y^*(k) \mathscr{Z}(k) \tag{4.47}$$

where

$$\mathscr{Z}(k) = [|z_1(k)|^2 z_1(k) \cdots |z_p(k)|^2 z_p(k)] \tag{4.48}$$

We now have to satisfy the orthogonality constraint at the next iteration of the algorithm:

$$\mathbf{G}^H(k+1)\mathbf{G}(k+1) = \mathbf{I} \tag{4.49}$$

For this to be feasible, we first assume that the channel output has been "prewhitened" in both "space" and time. This corresponds to assuming the channel matrix $\mathbf{H}$ to be unitary. Since any second-order blind separation method will typically converge to a unitary mixture of the inputs, this is a reasonable assumption: the MUK method can take over from there on to "unfold" blindly the unitary mixture.[3] Having assumed that $\mathbf{H}$ is unitary, to satisfy the constraint, Eq. (4.49), it suffices to satisfy

$$\mathbf{W}^H(k+1)\mathbf{W}(k+1) = \mathbf{I} \tag{4.50}$$

Because in general there is no guarantee that $\mathbf{W}'(k+1)$ will be unitary, we have to transform it to a unitary $\mathbf{W}(k+1) = f(\mathbf{W}'(k+1))$.

We propose choosing $\mathbf{W}(k+1)$ as an $m \times p$ matrix that is as close as possible to $\mathbf{W}'(k+1)$ in the Euclidean sense. Dropping for a while the

[3] Note that the required space–time prewhitening reduces to purely temporal prewhitening in the case of single-user equalization, as discussed in Section 4.3.

time index, this can be achieved with an iterative procedure that satisfies the following criterion successively for $j = 1, \ldots, p$:

$$\begin{cases} \min\limits_{W_j} & \Delta(W_j) = \|W_j - W_j'\|^2 \\ \text{subject to:} & W_l^H W_j = \delta_{lj}, \qquad l = 1, \ldots, j \end{cases} \tag{4.51}$$

where W_j' is defined from

$$\mathbf{W}' = [W_1' \cdots W_p'] \tag{4.52}$$

Problem (4.51) can also be written as

$$\begin{cases} \min\limits_{W_j} & \Delta(W_j) = (W_j - W_j')^H (W_j - W_j') \\ \text{subject to:} & W_l^H W_j = \delta_{lj}, \qquad l = 1, \ldots, j \end{cases} \tag{4.53}$$

To solve problem (4.53) we first construct the Lagrangian of $\Delta(W_j)$, which equals

$$\mathscr{L}_\Delta(W_j, \lambda_j, \boldsymbol{\mu}_j, \mathbf{v}_j) = \Delta(W_j) - \lambda_j(W_j^H W_j - 1) - \sum_{l=1}^{j-1} \mu_{lj} \operatorname{Re}(W_l^H W_j)$$
$$- \sum_{l=1}^{j-1} v_{lj} \operatorname{Im}(W_l^H W_j) \tag{4.54}$$

where λ_j, μ_{lj}, v_{lj} are real scalar parameters. Setting the gradient of $\mathscr{L}_\Delta(W_j, \lambda_j, \boldsymbol{\mu}_j, \mathbf{v}_j)$ with respect to W_j to zero $((\partial \mathscr{L}_\Delta / \partial W_j^*) = 0)$, we obtain the following equations for each j:

$$W_j - W_j' - \lambda_j W_j - \sum_{l=1}^{j-1} \beta_{lj} W_l = 0 \tag{4.55}$$

where $\beta_{lj} = 1/2(\mu_{lj} + \sqrt{-1} v_{lj})$. From Eq. (4.55), we obtain

$$\begin{cases} \lambda_j = 1 - W_j^H W_j' \\ \beta_{lj} = -W_l^H W_j' \end{cases} \tag{4.56}$$

which gives

$$(1 - \lambda_j) W_j = W_j' - \sum_{l=1}^{j-1} (W_l^H W_j') W_l \tag{4.57}$$

Table 4.1 The MUK Algorithm

1. $k = 0$: initialize $\mathbf{W}(0) = \mathbf{W}_0$
2. for $k > 0$
3. Obtain $\mathbf{W}'(k+1)$ from (4.47)
4. Obtain $W_1(k+1) = W_1'(k+1)/\|W_1'(k+1)\|$
5. for $j = 2 : p$
6. Compute $W_j(k+1)$ from $W_j'(k+1)$ through (4.59)
7. Go to 5
8. $\mathbf{W}(k+1) = [W_1(k+1) \cdots W_p(k+1)]$
9. Go to 2

According to Eq. (4.57),

$$W_j \propto \left(W_j' - \sum_{l=1}^{j-1} (W_l^H W_j') W_l \right) \tag{4.58}$$

where $\propto$ denotes "proportional." In light of Eq. (4.58), and keeping in mind that we must also satisfy $W_j^H W_j = 1$, we must choose W_j as

$$W_j(k+1) = \frac{W_j'(k+1) - \sum_{l=1}^{j-1} (W_l^H(k+1) W_j'(k+1)) W_l(k+1)}{\left\| W_j'(k+1) - \sum_{l=1}^{j-1} (W_l^H(k+1) W_j'(k+1)) W_l(k+1) \right\|} \tag{4.59}$$

(where we have brought back the time index k).

We summarize the MUK algorithm in Table 4.1. Notice that steps 4–8 correspond to a Gram-Schmidt orthogonalization of $\mathbf{W}'(k+1)$. We also note that the projection described by steps 4–7 of the algorithm would reduce to the mere normalization of step 4 in the case of a single user, as in Shalvi and Weinstein (1990) see (4.26). Also, if the transmitted input is nonsymmetrical, similar to Shalvi and Weinstein (1990), Eq. (4.47) should use the following vector instead of $\mathscr{Z}$ in Eq. (4.48):

$$\mathscr{Z}' = [|z_1|^2 z_1 - \widehat{z_1^2} z_1^* \cdots |z_p|^2 z_p - \widehat{z_p^2} z_p^*] \tag{4.60}$$

where $\hat{\cdot}$ again denotes empirical averaging.

Due to its constrained nature, the MUK algorithm requires, as mentioned earlier, an initial space-time pre-whitening of the received signals. This can be done with a number of different second-order BSS techniques, which typically converge to the required unitary mixture of signals.

As mentioned earlier, the MUK algorithm is very appealing from the point of view of convergence, namely, the following theorem can be shown:

Theorem 7. If all $\{a_i(k)\}, i = 1, \ldots, p$, share the same non-Gaussian pdf and are mutually independent and if H in Eq. (4.44) is a full column rank unitary matrix, then the MUK algorithm described in Table 4.1 is globally convergent to a solution that recovers perfectly (in the absence of noise) all the p input (source) signals (modulo a scalar phase rotation each). The proof of Theorem 7 relies on the particular (deflation-type) structure of the algorithm, as described in Eq. (4.59) (the proof of this theorem was derived only very recently and will be soon reported in the literature).

Other (non-MUK-type) single-stage constrained techniques have been proposed recently for the more specific problem of multiuser detection in CDMA systems. A maximum-output-energy (MOE) algorithm was proposed in Madhow et al. (1995) for multiuser detection in flat memoryless CDMA channels. This approach has an interesting duality to Minimum Variance Distortionless Response (MVDR) beam forming. Extensions to the case of unknown finite ISI were proposed more recently in Tsatsanis and Xu (1998) and Ghauri and Slock (1999).

In the next section we present some numerical examples of the performance of the MU-CMA and MUK algorithms described herein.

4.7 NUMERICAL EXAMPLES

4.7.1 MU-CMA

We consider a wireless communication scenario: two 4-QAM user signals (shaped by a raised-cosine pulse with roll-off parameter $\beta = 0.35$) impinge on a uniform linear array of $m = 10$ sensors, through two paths each (the delays and angles of arrival are $0.1T$, $2\pi/5$, $1.1T$, $-\pi/3$ and $0.4T$, $\pi/7$, $1.2T$, $-\pi/6$ for user 1 and 2, respectively), each with an SNR of 30 dB (here, T denotes the symbol period). The first user's signal arrives with a mean power that is 3 dB stronger than the second user's signal. The multiple-input-multiple-output (MIMO) equalizer is a 20×2 matrix ($M = 2$) and is initialized with only two nonzero elements, equal to 1, at the positions $(10, 1)$ and $(12, 2)$. Figure 4.4 shows the performance of the MU-CMA in terms of output mean-squared error (MSE). The MSE is calculated by sample averaging each output's squared error (after the arbitrary rotation induced by the algorithm is removed) over a

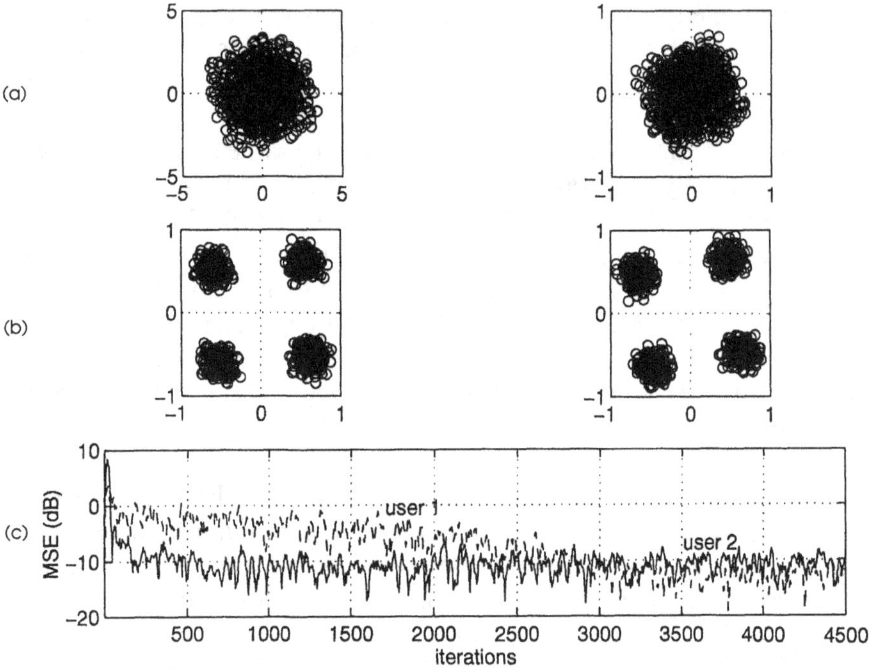

Figure 4.4 Performance of the MU-CMA on a 2×10 convolutive-mixture channel.

30-point rectangular window. The step-size $\mu = 0.0005$, $\alpha = 2$, and we choose $\delta_1 = -\delta_2 = -2$. The upper plots in Fig. 4.4*a* show the channel eye at two randomly chosen (out of the 10) antenna elements before the algorithm is run. Notice how the channel eye is initially completely closed, corresponding to very poor performance. The two lower plots in Fig. 4.4*b* show the corresponding channel eye of each of the two outputs of the system after the algorithm's convergence. Notice that both channel eyes are now open, corresponding to a proper recovery of the two transmitted signals. Figure 4.4*c* shows the evolution of the MSE of each of the two signals (after the removal of the induced arbitrary phase rotation). Notice that the first user signal is recovered faster (which is probably due to the fact that it is stronger in power). Also notice that both signals are recovered at about a 20-dB squared error better than the corresponding initial values.

4.7.2 MUK Algorithm

We next present some examples of the performance of the MUK algorithm described in Table 4.1. We first consider the simple case of a 2×2

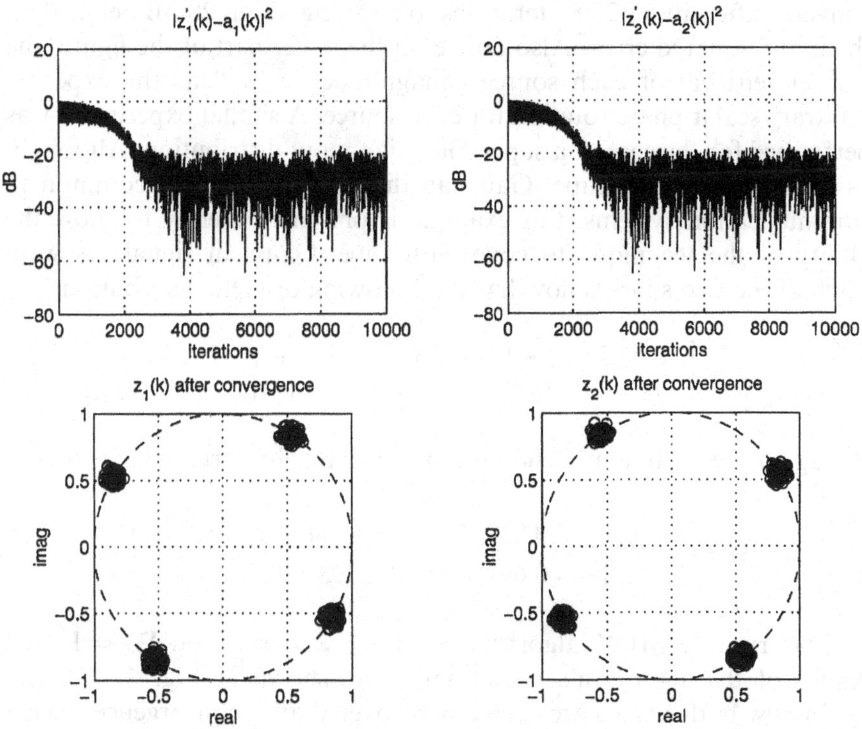

Figure 4.5 Performance of the MUK algorithm for $p = 2$ 4-QAM signals.

unitary matrix channel chosen randomly as:

$$\mathbf{H} = \begin{bmatrix} 0.701 + 0.172\sqrt{-1} & 0.629 + 0.286\sqrt{-1} \\ -0.274 - 0.634\sqrt{-1} & 0.159 + 0.704\sqrt{-1} \end{bmatrix}$$

We then run the MUK algorithm with a step size $\mu = 2 \times 10^{-3}$ and with $\mathbf{W}_0 = \mathbf{I}$ for the case of two independent 4-QAM inputs that go through the channel, and are received with a 30-dB SNR at the channel output. After convergence, the algorithm reaches the following setting:

$$\mathbf{G} = \begin{bmatrix} 0.974 + 0.222\sqrt{-1} & 0.028 + 0.014\sqrt{-1} \\ -0.013 - 0.028\sqrt{-1} & 0.204 + 0.978\sqrt{-1} \end{bmatrix}$$

which corresponds to the recovery of the first source from the first output, and the second source from the second output. Figure 4.5 shows the evolution of the instantaneous squared error in decibels between each equalizer output (after phase correction) and the corresponding retrieved source, as the algorithm progresses. Notice how both channel outputs converge after about 2000 iterations to a setting about 20 dB better than the initial squared error. Also notice, in the lower part of the figure, the correct retrieval of each source's magnitude, as well as the expected arbitrary scalar phase rotation for each source. A similar experiment was performed for the case of a super-Gaussian input distribution. Although, as mentioned above, super-Gaussian distributions are not common in communication systems, this example is presented merely to show the ability of the technique to cope with super-Gaussian signals as well. Each of the two sources now has the following discrete distribution:

$$\begin{bmatrix} a_j \\ \Pr(a_j) \end{bmatrix} = \begin{bmatrix} 0 & 1+\sqrt{-1} & 1-\sqrt{-1} & -1+\sqrt{-1} & -1-\sqrt{-1} \\ 3/4 & 1/16 & 1/16 & 1/16 & 1/16 \end{bmatrix}$$

We again use a randomly chosen unitary channel matrix, which equals

$$\mathbf{H} = \begin{bmatrix} -0.307+0.071\sqrt{-1} & -0.844+0.379\sqrt{-1} \\ -0.616+0.691\sqrt{-1} & 0.1798-0.3315\sqrt{-1} \end{bmatrix}$$

and we run the MUK algorithm with $\mu = 2 \times 10^{-3}$ and $\mathbf{W}_0 = \mathbf{I}$. The results of the algorithm's performance are shown in Fig. 4.6. Notice again how both signals are correctly recovered after convergence (a case in which CM-type algorithms would fail). Notice also that the first receiver output recovers the second user's input and vice versa, reflecting the mentioned ambiguity in ordering, which is inherent to blind multiuser criteria.

In a fourth experiment, we consider $p = 4$ independent 4-QAM sources. We then run 100 independent runs, where the signals are mixed through a different randomly chosen 4×4 unitary matrix each time (again the SNR is fixed at 30 dB and $\mu = q \times 10^{-3}$). In Fig. 4.7 we show the squared error between each channel output and the (corresponding) retrieved source, averaged over the 100 independent runs. Notice that the average plots show the correct retrieval of the four sources, over a large number of channel realizations, reflecting the expected (according to Theorem 7) globally-convergent behavior of the algorithm for an arbitrary number of input sources. This also shows that the algorithm's convergence is robust with respect to the conditioning of the channel matrix. Similar behavior has been verified for channels of differrent dimensions.

4.8 CONCLUSIONS

In this chapter we have studied the problem of blind separation of independent sources from an identifiability perspective. We have consid-

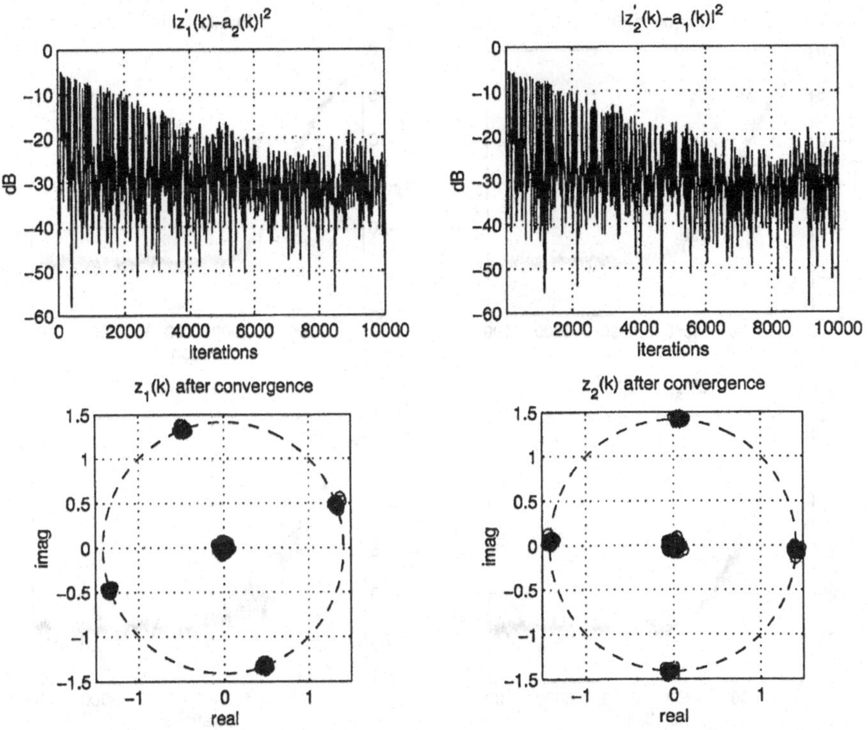

Figure 4.6 Performance of the MUK algorithm for $p = 2$ super-Gaussian signals.

ered a generic system setup wherein a number of user (source) signals are transmitted through a MIMO linear dispersive channel that introduces both ISI and IUI. Motivated by the identifiability results obtained by Shalvi and Weinstein (1990) in the single-user BE case, we have examined the existence of necesseary and sufficient conditions for the recovery of all the transmitted signals. We assumed that all the sources share the same non-Gaussian distribution and that they are individually i.i.d. and mutually independent. Based on these assumptions, we presented a set of conditions that are necessary and sufficient for the recovery of all the sources. This set of conditions can then be used as a practical guideline for the design of BSS criteria. Such criteria can be broadly classified in two distinct categories: constrained and unconstrained criteria. We have described some of the techniques that fall into these categories, and have focused on one representative approach for each category. We presented gradient-type algorithms for these criteria and discussed their performance based on both theoretical analysis and computer-simulated results. As expected by their agreement with the necessary and sufficient identifiability conditions, and by virtue of their relatively low computa-

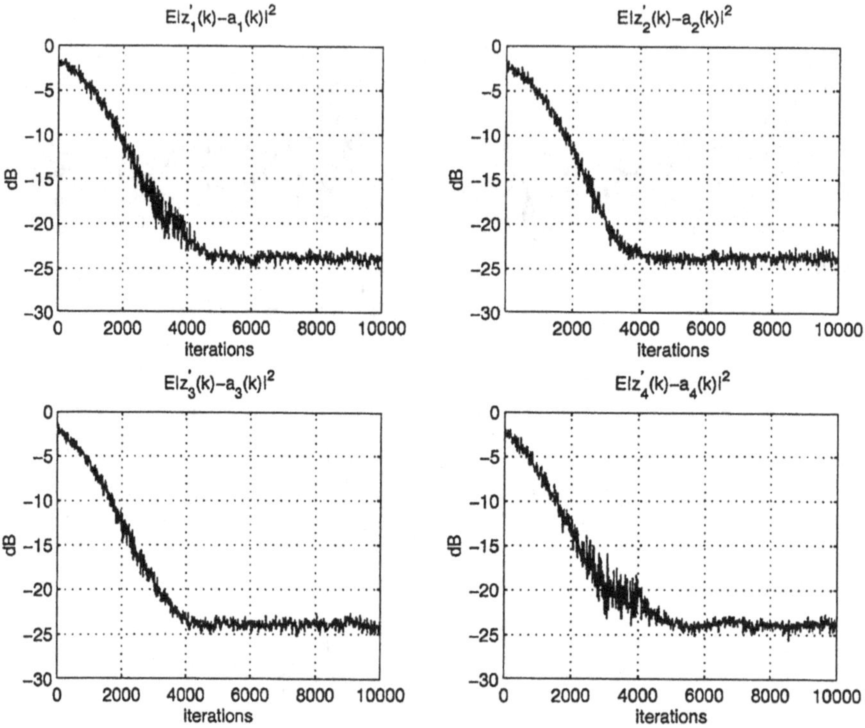

Figure 4.7 Average performance of the MUK algorithm for $p = 4$ on 100 random channels.

tional complexity and good convergence properties, it is our belief that these techniques show significant promise for successful BSS in practical applications.

REFERENCES

Benveniste, A., M. Goursat, and G. Ruget, June 1980, "Robust identification of a nonminimum phase system: blind adjustment of a linear equalizer in data communications," *IEEE Trans. Automatic Control*, vol. AC-25, no. 3, pp. 1740–1748.

Cardoso, J. F., 1989, "Source separation using higher order moments," in *IEEE Int. Conf. on Acoustics, Speech, and Signal Processing*, vol. 4, Adelaide, Australia, pp. 2109–2112.

Cardoso, J. F., and B. H. Laheld, Dec. 1996, "Equivariant adaptive source separation," *IEEE Trans. Signal Processing*, vol. 44, no. 12, pp. 3017–3030.

Castedo, L., C. J. Escudero, and A. Dapena, June 1997, "A blind signal separation method for multiuser communications," *IEEE Trans. Signal Processing*, vol. 45, pp. 1343–1348.

Comon, P., April 1994, "Independent components analysis: a new concept?," *Signal Processing*, vol. 36, no. 3, pp. 287–314.

Comon, P., July 1996, "Contrasts for multichannel blind deconvolution," *IEEE Signal Processing Lett.*, vol. 3, no. 7, pp. 209–211.

Delfosse, N., and P. Loubaton, April 1995, "Adaptive separation of independent sources: A deflation approach," in *IEEE Int. Conf. on Acoustics, Speech, and Signal Processing*, Adelaide, Australia, pp. 41–44.

Donoho, D., 1981, "On minimum entropy deconvolution," *Applied Time-Series Analysis II*, (New York: Academic Press), pp. 565–609.

Fijalkow, I., and P. Loubaton, 1995, "Identification of rank one rational spectral densities from noisy observations: a stochastic realization approach," *Systems and Control Lett.*, vol. 24, pp. 201–205.

Fijalkow, I., A. Touzni, and J. R. Treichler, Jan. 1997, "Fractionally-spaced equalization using CMA: robustness to channel noise and lack of disparity," *IEEE Trans. on Signal Processing, Special Issue on Signal Processing for Advanced Communications*, vol. 45, no. 1, pp. 56–66.

Foschini, G. J., Oct. 1985, "Equalizing without altering or detecting data," *AT&T Tech. J.*, vol. 64, no. 8, pp. 1885–1911.

Ghauri, I., and D. T. M. Slock, 1999, "Blind channel and linear MMSE receiver determination in DS-CDMA systems," *Proc. ICASSP 99 Conference*, Arizona, USA, March 15–19, 1999.

Godard, D. N., Nov. 1980, "Self-recovering equalization and carrier tracking in two-dimensional data communication systems," *IEEE Trans. Communications*, vol. COM-28, no. 11, pp. 1867–1875.

Gooch, R. P., and J. Lundel, 1986, "The CM array: an adaptive beamformer for constant modulus signals," in *Int. Conf. on Acoustics, Speech, and Signal Processing*, pp. 2523–2526.

Inouye, Y., 1997, "Blind deconvolution of multichannel linear time-invariant systems of nonminimum phase, in T. Katayama and S. Sugimoto, eds., *Statistical Methods in Control and Signal Processing*," (New York: Marcel Dekker), pp. 375–397.

Inouye, Y., Dec. 1998, "Criteria for blind deconvolution of multichannel linear time-invariant systems," *IEEE Trans. Signal Processing*, vol. 46, no. 12, pp. 3432–3436.

Lambotharan, S., and J. Chambers, April 1999, "On the surface characteristics of a mixed constant modulus and cross-correlation criterion for the blind equalization of a MIMO channel," *Signal Processing*, vol. 74, pp. 209–216.

Li, Y., and Z. Ding, April 1996, "Global convergence of fractionally spaced Godard (CMA) adaptive equalizers," *IEEE Trans. Signal Processing*, vol. 44, no. 4, pp. 818–826.

Madhow, U., M. L. Honig, and S. Verdú, July 1995, "Blind adaptive multiuser detection," *IEEE Trans. Inform. Theory*, vol. 41, pp. 944–960.

Papadias, C. B., 1998, "Kurtosis-based criteria for adaptive blind source separation," in *IEEE Int. Conf. Acoust., Speech, and Signal Processing*, Seattle, WA, USA, May 12–15, 1998, pp. 2317–2320.

Papadias, C. B., Feb. 1999, "Blind separation of independent sources based on a multi-user kurtosis maximization criterion," *Bell Labs Technical Memorandum*.

Papadias, C. B., and A. Paulraj, June 1997a, "A constant modulus algorithm for multi-user signal separation in presence of delay spread using antenna arrays," *IEEE Signal Processing Lett.*, vol. 4, no. 6, pp. 178–181.

Papadias, C. B., and A. J. Paulraj, 1997b, "Blind separation of independent cochannel signals," in *13th Int. Conf. on Digital Signal Processing*, Santorini, Greece, July 2–4, 1997, pp. 139–142.

Papadias, C. B., and D. T. M. Slock, March 1999, "Fractionally spaced equalization of linear polyphase channels and related blind techniques based on multichannel linear prediction," *IEEE Trans. Signal Processing*, vol. 47, no. 3, pp. 641–654.

Paulraj, A., C. B. Papadias, V. U. Reddy, and A.-J. van der Veen, 1997, "Space-time blind signal processing for wireless communication systems: recent advances and practical considerations," in G. Wornell and H. V. Poor, eds., *Wireless Communications: A Signal Processing Perspective*," (Englewood Cliffs, NJ: Prentice Hall), pp. 179–210.

Romano, J. M. T., F. R. P. Cavalcanti, and A. L. Brandao, 1999, "Least-squares CMA with decorrelation for fast blind multiuser signal separation," *Proc. ICASSP 99 Conference*, Arizona, USA, March 15–19, 1999.

Shalvi, O., and E. Weinstein, March 1990, "New criteria for blind deconvolution of non-minimum phase systems," *IEEE Trans. Inform. Theory*, vol. 36, pp. 312–321.

Shamsunder, S., and G. Giannakis, 1994, "Multichannel blind signal separation and reconstruction," in *6th IEEE Digital Signal Processing Workshop*, Yosemite, CA.

Shynk, J. J., and R. P. Gooch, March 1996, "The constant modulus array for cochannel signal copy and direction finding," *IEEE Trans. Signal Processing*, vol. 44, no. 3, pp. 652–660.

Slock, D. T. M., 1994, "Blind joint equalization of multiple synchronous mobile users using oversampling and/or multiple antennas," in *28th Asilomar Conf. on Signals, Systems, and Computers*, Oct. 31–Nov. 2, Pacific Grove, CA, pp.

Swami, A., G. Giannakis, and S. Shamsunder, April 1994, "Multichannel ARMA processes," *IEEE Trans. Signal Processing*, vol. 42, no. 4, pp. 898–913.

Touzni, A., and I. Fijalkow, 1997, "Blind adaptive equalization and simultaneous separation of convolutive mixtures," in *13th Int. Conf. on Digital Signal Processing*, Santorini, Greece, July 2–4, 1997, pp. 391–394.

Treichler, J. R., and B. G. Agee, April 1983, "A new approach to multipath correction of constant modulus signals," *IEEE Trans. Acoustics, Speech, and Signal Processing*, vol. ASSP-31, no. 2, pp. 459–472.

Treichler, J. R., and M. G. Larimore, April 1985, "New processing techniques based on the constant modulus adaptive algorithm," *IEEE Trans. Acoust., Speech, and Signal Processing*, vol. ASSP-33, no. 2, pp. 420–431.

Tsatsanis, M. K., and Z. (Daniel) Xu, Nov. 1998, "Performance analysis of minimum variance CDMA receivers," *IEEE Trans. Signal Processing*, vol. 46, no. 11, pp. 3014–3022.

Tugnait, J. K., Jan. 1997, "Blind spatio-temporal equalization and impulse response estimation for MIMO channel using a Godard cost function," *IEEE Trans. on Signal Processing, Special Issue on Signal Processing for Advanced Communications*, vol. 45, no. 1, pp. 268–271.

van der Veen, A., and A. Paulraj, May 1996, "An analytical constant modulus algorithm," *IEEE Trans. Signal Processing*, vol. 44, pp. 1136–1155.

Yellin, D., and E. Weinstein, Aug. 1994, "Criteria for multichannel signal separation," *IEEE Trans. Signal Processing*, vol. 42, no. 8, pp. 2158–2168.

INDEX

9 780471 379416